COMMON COR
MATH 4
WORKBOOK

MW00388089

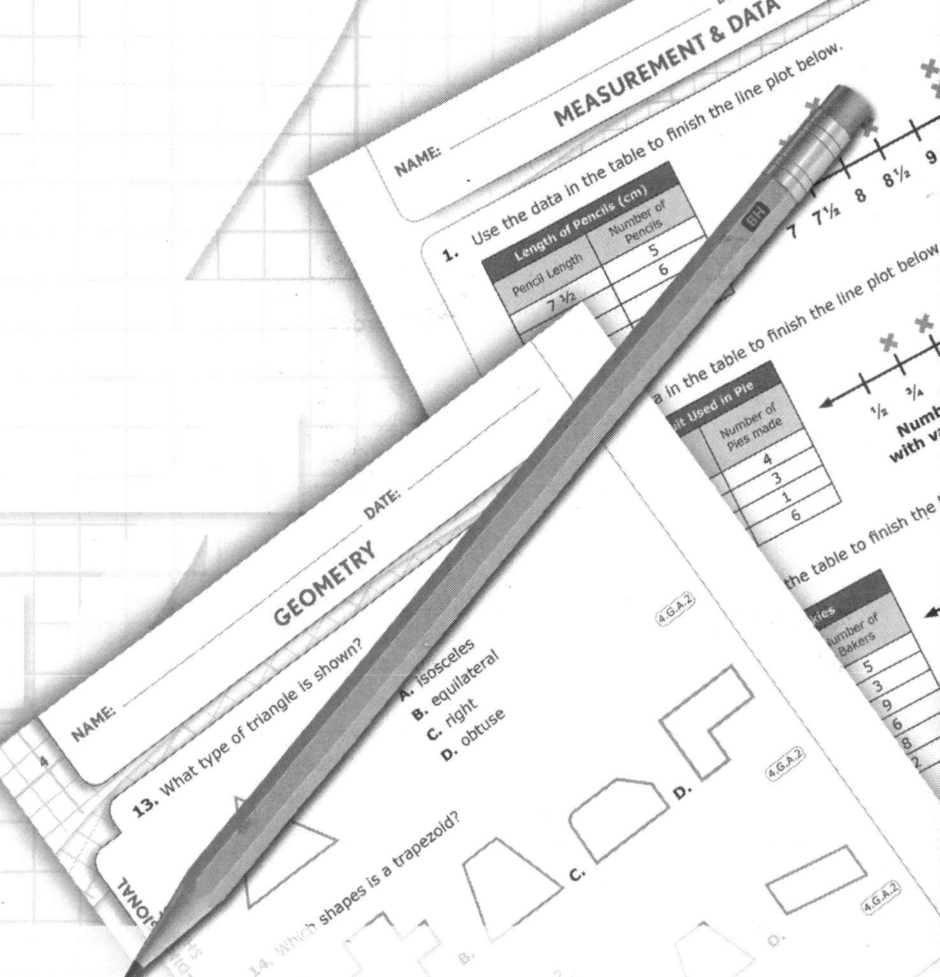

Ace Academic Publishing
ACHIEVING EXCELLENCE TOGETHER

www.aceacademicprep.com

Author: Ace Academic Publishing

Ace Academic Publishing is a leading supplemental educational provider committed to offering students an enjoyable and interactive learning experience. Through our comprehensive workbooks that are designed to include challenging, multi-step questions, we aim to provide students with state of the art educational materials that will help them improve their academic performance. Our carefully selected practice questions encourage logical thinking and creativity and combine the focus on the required common core standards along with the understanding of the practical applications of the mathematical concepts.

For inquiries, contact Ace Academic Publishing at the following address:

Ace Academic Publishing
3736 Fallon Road #403
Dublin CA 94568

www.aceacademicprep.com

This book contains copyright protected material. The purchase of this material entitles the buyer to use this material for personal and classroom use only. Reproducing the content for commercial use is strictly prohibited. Contact us to learn about options to use it for an entire school district or other commercial use.

ISBN:978-1-949383-03-4

© Ace Academic Publishing, 2018

About the Book

The contents of this book includes multiple chapters and units covering all the required Common Core Standards for this grade level. Similar to a standardized exam, you can find questions of all types, including multiple choice, fill-in-the-blank, true or false, matching and free response questions. These carefully written questions aim to help students reason abstractly and quantitatively using various models, strategies, and problem-solving techniques. The detailed answer explanations in the back of the book help the students understand the topics and gain confidence in solving similar problems.

For the Parents

This workbook includes practice questions and tests that cover all the required Common Core Standards for the grade level. The book is comprised of multiple tests for each topic so that your child can have an abundant amount of tests on the same topic. The workbook is divided into chapters and units so that you can choose the topics that you want your child needs to focus on. The detailed answer explanations in the back will teach your child the right methods to solve the problems for all types of questions, including the free-response questions. After completing the tests on all the chapters, your child can take any Common Core standardized exam with confidence and can excel in it.

For additional online practice, sign up for a free account at www.aceacademicprep.com.

For the Teachers

All questions and tests included in this workbook are based on the Common Core State Standards and includes a clear label of each standard name. You can assign your students tests on a particular unit in each chapter, and can also assign a chapter review test. The book also includes two final exams which you can use towards the end of the school year to review all the topics that were covered. This workbook will help your students overcome any deficiencies in their understanding of critical concepts and will also help you identify the specific topics that your students may require additional practice. These grade-appropriate, yet challenging, questions will help your students learn to strategically use appropriate tools and excel in Common Core standardized exams.

For additional online practice, sign up for a free account at www.aceacademicprep.com.

For additional practice and help, visit our website at www.aceacademicprep.com

You can find more workbooks for Math and English for all grade levels

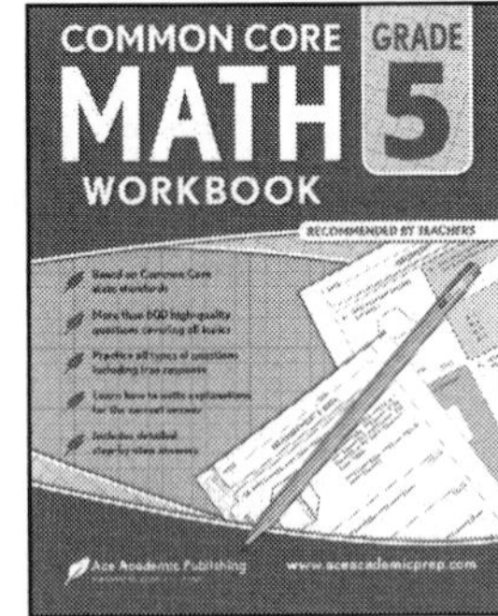

Ace Academic Publishing
ACHIEVING EXCELLENCE TOGETHER

www.aceacademicprep.com

TABLE OF CONTENTS

OPERATIONS & ALGEBRAIC THINKING

PROBLEM SOLVING-ALL OPERATIONS 9
FACTORS AND MULTIPLES 14
PATTERNS 18
CHAPTER REVIEW 22
EXTRA PRACTICE 26

NUMBER & OPERATIONS IN BASE TEN

PLACE VALUE - ROUNDING AND COMPARING 33
ADDITION AND SUBTRACTION 39
MULTIPLICATION AND DIVISION 46
CHAPTER REVIEW 51
EXTRA PRACTICE 58

NUMBER & OPERATIONS - FRACTIONS

EQUIVALENT FRACTIONS AND ORDERING FRACTIONS 69
FRACTION OPERATIONS 75
DECIMAL CONVERSION AND COMPARISON 82
CHAPTER REVIEW 89
EXTRA PRACTICE 96

MEASUREMENT & DATA

PROBLEM SOLVING - CONVERSION OF MEASUREMENTS 107
GRAPHS AND DATA INTERPRETATION 111
GEOMETRY - CONCEPTS OF ANGLES 121
CHAPTER REVIEW 128
EXTRA PRACTICE 134

GEOMETRY

LINES AND ANGLES 143
TWO-DIMENSIONAL SHAPES 149
LINE OF SYMMETRY 154
CHAPTER REVIEW 158
EXTRA PRACTICE 162

COMPREHENSIVE ASSESSMENT 1 171

COMPREHENSIVE ASSESSMENT 2 190

ANSWERS AND EXPLANATIONS 205

OPERATIONS & ALGEBRAIC THINKING

PROBLEM SOLVING-ALL OPERATIONS 9

FACTORS AND MULTIPLES 14

PATTERNS 18

CHAPTER REVIEW 22

EXTRA PRACTICE 26

1. An orchard has 15 bags of small apples. They have 6 times as many bags of large apples as small apples. Which equation tells how many bags of large apples the orchard has?

A. $15 \times 6 = 90$ **B.** $15 \times 90 = 6$

C. $90 \times 6 = 15$ **D.** $6 \times 90 = 15$

(4.OA.A.1)

2. Which equation best matches this story?

Robert has $14. Julie has twice as much money as Robert. Let the variable x represent Julie's money. How much money does Julie have?

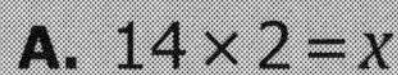

A. $14 \times 2 = x$ **B.** $x \times 2 = 14$

C. $2 \times x = 14$ **D.** $14 + 2 = x$

(4.OA.A.1)

3. Which comparison sentence best describes the equation?

A. The number 6 is 12 times as many as 72.

B. The number 12 is 6 times as many as 72.

C. The number 12 is 72 times as many as 6.

D. The number 72 is 6 times as many as 12.

(4.OA.A.1)

4. Volunteers at the school bake sale are selling cupcakes and cookies. They sold seven times as many cookies as cupcakes and they sold nine cupcakes. Which equation tells how many cookies they sold?

A. $9 = 7 \times x$ **B.** $x + 7 = 9$

C. $7 = x \times 9$ **D.** $x = 7 \times 9$

(4.OA.A.1)

PROBLEM SOLVING ALL OPERATIONS

5. Which story is represented by the equation $4 \times 3 = t$?

A. A toy car costs $12. The toy robot costs 3 times the cost of the toy car. How much does the robot cost?

B. A toy car costs $4. The toy robot costs 12 times the cost of the toy car. How much does the robot cost?

C. A toy car costs $4. The toy robot costs 3 times the cost of the toy car. How much does the robot cost?

D. A toy car costs $4. The toy robot costs $3. How much would it cost to buy both the car and the robot?

4.OA.A.1

6. Jessica says that $? = 8 \times 6$ is equal to $? = 6 \times 8$. Do you agree or disagree? State your reasoning.

____

____

____

____

4.OA.A.1

7. Joseph is trying to write an equation to go with the statement, "45 is 5 times as many as 9." What is the equation?

____

____

____

4.OA.A.1

PROBLEM SOLVING ALL OPERATIONS

8. Jacob is thinking of a number that is 13 times the number 6. What number is he thinking of? How do you know?

____

____

____

4.OA.A.2

9. John picked 65 red flowers, which is 5 times as many as the number of yellow flowers he picked. How many yellow flowers did he pick? Explain your reasoning.

____

____

____

____

4.OA.A.2

10. You are helping your little brother with his math homework. He wants to find 15 times 5 but he is not sure how. Explain to him one strategy he could use.

____

____

____

____

4.OA.A.2

PROBLEM SOLVING ALL OPERATIONS

11. There are 21 kindergarteners in each class. Four kindergarten classes are lined up waiting for lunch. Which equation shows the number of kindergarteners who are waiting for lunch?

A. $21 + 4 = k$

B. $21 \times k = 4$

C. $21 \times 4 = k$

D. $4 \times k = 21$

(4.OA.A.2)

12. In the first quarter of the basketball game, 12 points were scored. In the second quarter, 4 times as many points were scored than in the first quarter. Which equation shows how many points were scored in the second quarter?

A. $12 + 4 = ?$

B. $? - 4 = 12$

C. $12 \times ? = 4$

D. $12 \times 4 = ?$

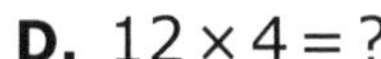

(4.OA.A.2)

13. Each of the 60 students in fifth grade need 4 colored pencils for a project. How many pencils are needed altogether?

____

(4.OA.A.2)

14. A bakery has an order for 6 times as many strawberry cupcakes as chocolate cupcakes. If the order includes 7 chocolate cupcakes, how many strawberry cupcakes were ordered?

____

(4.OA.A.2)

15. Stephanie gets $5.00 each week for her allowance. She has saved all her allowance for the last 17 weeks. She buys 2 games that cost a total of $80. How much money does she have left?

A. $5.00 **B.** $10.00 **C.** $15.00 **D.** $20.00

(4.OA.A.3)

16. A shop has 53 medium shirts and 41 large shirts to sell. If the workers pack the shirts into boxes with exactly 10 shirts in each box, how many shirts will be left over?

A. 0 **B.** 4 **C.** 5 **D.** 1

(4.OA.A.3)

17. Each of the 30 students in Mr. Anderson's class completed a sheet of math problems that has 5 problems on the front and 6 problems on the back. Which answer is the best estimation of the total number of problems the class has completed?

A. A little less than 300
B. A little more than 300
C. A little less than 200
D. A little more than 200

(4.OA.A.3)

18. James has counted 60 spots on ladybugs in the field, then he counted 60 spots on the ladybugs by his house. If each ladybug has 6 spots, how many ladybugs did he count?

A. 10 **B.** 20 **C.** 19 **D.** 22

(4.OA.A.3)

19. Samantha wants to buy a toy that costs $18.00. Her Grandpa gave her and her three siblings $60 to share equally. How much more money does she need to buy the toy?

____

(4.OA.A.3)

20. There are 82 books in Ms. Vegas' classroom library. If each of the 31 students in Ms. Vega's class takes 2 books how many books will be left?

____

(4.OA.A.3)

OPERATIONS & ALGEBRAIC THINKING

FACTORS AND MULTIPLES

1. Which of the following numbers is not a composite number?

A. 61 **B.** 20 **C.** 33 **D.** 77

4.OA.B.4

2. What is the next prime number after 67?

A. 73 **B.** 69 **C.** 71 **D.** 79

4.OA.B.4

3. Which of the following numbers is not prime?

A. 2 **B.** 41 **C.** 17 **D.** 21

4.OA.B.4

4. How many factor pairs does the number 22 have?

A. 0 **B.** 1 **C.** 2 **D.** 3

4.OA.B.4

5. Which number is not a factor of 44?

A. 44 **B.** 11 **C.** 1 **D.** 6

4.OA.B.4

6. Which number is a factor of 98?

A. 45 **B.** 49 **C.** 3 **D.** 4

4.OA.B.4

7. How many factor pairs does the number 88 have?

A. 1 **B.** 2 **C.** 3 **D.** 4

4.OA.B.4

8. Which number has 4 as one of its factors?

A. 78 **B.** 82 **C.** 70 **D.** 72

4.OA.B.4

9. Which number does not have 3 as one of its factors?

A. 90 **B.** 86 **C.** 96 **D.** 57

4.OA.B.4

10. Which number has 6 as one of its factors?

A. 82 **B.** 91 **C.** 96 **D.** 86

4.OA.B.4

11. Is 37 a prime or a composite number? How do you know?

____

____

____

4.OA.B.4

12. What is the next prime number after 17? How do you know?

____

____

____

4.OA.B.4

FACTORS AND MULTIPLES

13. What are all the factor pairs of 62? How do you know?

____

____

____

4.OA.B.4

14. Is 3 a factor of 94?

____

4.OA.B.4

15. A number that has more than one factor pair is called

a ____________ number.

4.OA.B.4

16. How many prime numbers are between 1 and 10?

____

4.OA.B.4

17. What is the next composite number after 66?

____

4.OA.B.4

18. Is the number 89 prime or composite?

____

4.OA.B.4

19. A number that has exactly two factors is called a ________________ number.

(4.OA.B.4)

20. How many prime numbers are there between 70 and 80?

____

(4.OA.B.4)

UNIT 3: PATTERNS

1. Max trims 3 trees an hour. What number is missing in the table?

Hours	Trees Trimmed
1	3
10	30
100	?

A. 90
B. 350
C. 300
D. 60

4.OA.C.5

2. What is the rule of the pattern?

A. Quarter turn
B. Half turn
C. Flip
D. Slide

4.OA.C.5

3. What is the next number in the series 2, 4, 8, 16?

A. 20
B. 30
C. 26
D. 32

4.OA.C.5

4. What is true about a number pattern that starts at 300 with a rule of "subtract 5"?

A. Numbers in the pattern will be even.
B. Numbers in the pattern will be odd.
C. Numbers in the pattern will alternate between even and odd.
D. Numbers in the pattern will be even twice, then odd once.

4.OA.C.5

PATTERNS

5. What will be true about a number pattern that is starts at 222 with a rule of "add 2"?

A. Numbers in the pattern will be even.
B. Numbers in the pattern will be odd.
C. Numbers in the pattern will alternate between even and odd.
D. Numbers in the pattern will be even twice, then odd once.

4.OA.C.5

6. What is the next number in this pattern?

9, 18, 27... .

A. 34 **B.** 36 **C.** 45 **D.** 39

4.OA.C.5

7. What is true about a pattern that starts at 23 with a rule of "add 3"?

A. Numbers in the pattern will be even.
B. Numbers in the pattern will be odd.
C. Numbers in the pattern will alternate between even and odd.
D. Numbers in the pattern will be even twice then odd once.

4.OA.C.5

8. What number goes in the blank to complete the pattern?

16, ________, 24, 28

4.OA.C.5

9. What number goes in the blank to complete the pattern?

71, ________, 81, 86

4.OA.C.5

PATTERNS

10. What is the rule of the sequence 48, 60, 72 ... .?

The rule is ________________.

(4.OA.C.5)

11. The first number in the sequence is 64. The rule is "add 10". What is the fourth number in the sequence?

____

(4.OA.C.5)

12. What number is missing from the sequence?

1, 5, ________, 125, 625

(4.OA.C.5)

13. Part of a sequence is ... 75, 90, 105...

What number goes directly before 75 in the sequence?

____

(4.OA.C.5)

14. The first number in a sequence is 55. The rule is "multiply by 2 then subtract 10". What number will be third in the sequence?

____

(4.OA.C.5)

15. What number is missing from this sequence?

....160, ________, 40, 20...

(4.OA.C.5)

16. The first number in a sequence is 85. The rule is "subtract 12". What is the third number in the sequence?

____

(4.OA.C.5)

17. The third number in a sequence is 78. The rule is "add 11". What is the first number in the sequence?

____

(4.OA.C.5)

18. If I start with 4 and make a pattern with the rule "add 4", will the numbers in my pattern be even, odd or a combination of both? How do you know?

____

____

(4.OA.C.5)

PATTERNS

19. What rule is used to create this pattern? Explain how you know.

4, 8, 12, 16

____

____

(4.OA.C.5)

20. What shape is the next figure in the pattern? Why?

____

____

____

(4.OA.C.5)

CHAPTER REVIEW

CHAPTER REVIEW

1. Which story does the equation best match? The variable *x* represents Kate's age.

A. Kate's uncle is 9 times as old as she is. Her uncle is 9. How old is Kate?

B. Kate's uncle is 9 times as old as she is. Her uncle is 36. How old is Kate?

C. Kate's age plus her uncle's age equals 36. How old is Kate?

D. Kate's 9-year-old uncle is 36 times as old as she is. How old is Kate?

4.OA.A.1

2. Matt has three times as many shirts as Pat has. Pat has 4 t-shirts. Which equation shows the number of shirts Matt has? The variable represents the number of shirts Matt has.

A. $4 \times 3 = s$ **B.** $S \times 3 = 4$ **C.** $4 + 3 = s$ **D.** $4 \times s = 3$

4.OA.A.1

3. The equation describes the comparison ________ is 14 times 6. What number goes in the blank?

A. 8 **B.** 20 **C.** 14 **D.** 84

4.OA.A.1

4. The comparison 63 is 7 times 9 is represented by the equation 63 = 9 × ________. What number goes in the blank?

A. 63 **B.** 9 **C.** 7 **D.** 54

4.OA.A.1

5. How many factor pairs does 50 have?

A. 6 **B.** 3 **C.** 4 **D.** 5

(4.OA.B.4)

6. The number 2 and what other number make a factor pair of 58?

A. 29 **B.** 22 **C.** 28 **D.** 30

(4.OA.B.4)

7. Fill in the Blank.

The factors of a prime number are ________ and the number itself.

(4.OA.B.4)

8. Is the number 91 prime or composite?

____

(4.OA.B.4)

9. I read 45 book chapters this month. That is 3 times the number of books that I read last month. How many chapters did I read last month?

A. 90 **B.** 135 **C.** 15 **D.** 13

(4.OA.A.2)

10. A bowl with 15 pieces of fruit is sitting on each table in a dining room. There are 9 tables in the room. How many pieces of fruit are in the dining room?

A. 130 **B.** 135 **C.** 24 **D.** 140

(4.OA.A.2)

CHAPTER REVIEW

11. Marcus and Jessica are practicing free throws. Marcus has made 8 baskets so far. Jessica has made 3 times as many baskets as Marcus. How many baskets has Jessica made?

A. 25 **B.** 24 **C.** 12 **D.** 26

4.OA.A.2

12. Each of the 27 students in Mr. Park's class can check out up to 5 books. What is the maximum number of books in Mr. Park's class available for students to check out?

____

4.OA.A.2

13. What is the next number in the pattern?

350, 315, 280, ...

A. 240 **B.** 245 **C.** 250 **D.** 235

4.OA.C.5

14. James wrote the following pattern:

22, 33, 44...

What is the rule of his pattern?

A. Add 10 **B.** Add 11 **C.** Multiply by 2 **D.** Add 12

4.OA.C.5

15. Part of a sequence is ... 64, 128, 256 ...

What number comes directly before 64 in the sequence?

A. 612 **B.** 32 **C.** 20 **D.** 28

4.OA.C.5

16. What is the next number in this pattern?

81, 27, 9 ... . ?

____

(4.OA.C.5)

17. Ms. Jacobs is giving away 45 books that she no longer needs. She saves 10 books for the students in Ms. Thompson's class, and then divides the rest equally among 7 students. How many books will each of the 7 students receive?

A. 4 **B.** 5 **C.** 6 **D.** 7

(4.OA.A.3)

18. Pedro is thinking of a number that is 24 more than 75 divided by 5. What number is Pedro thinking of?

____

(4.OA.A.3)

19. A fisherman caught 42 fish on his first trip and 38 fish on his second trip. He is going to put the fish into packs of 6. How many full packs will he be able to make?

____

(4.OA.A.3)

20. A bakery has 130 cupcakes. They make 5 packs of 12 cupcakes and then will sell the rest one at a time. How many cupcakes will not be packaged?

____

(4.OA.A.3)

EXTRA PRACTICE

EXTRA PRACTICE

1. The number 90 is 10 times 9.

Write this comparison as a multiplication equation.

____

(4.OA.A.1)

2. I have 8 bags of marbles. Each bag has 11 marbles in it. I have 88 marbles. Write a multiplication equation that matches this story.

____

(4.OA.A.1)

3. Six students are playing in the field. Four times as many students are playing on the playground than in the field. How many students are on the playground? Write an equation to represent this story. Use to represent the unknown number.

____

(4.OA.A.1)

4. The number 90 is 15 times number 6. Write a multiplication equation that matches this comparison.

____

(4.OA.A.1)

5. Fill in the Blank. The equation $20 = 4 \times 5$ shows that 20 is 4 times 5.

It could also show that 20 is ________ times 4.

(4.OA.B.4)

6. James is 7 times as old as Jane. Jane is 4 years old. Write an equation to represent this story. Use to represent the unknown number.

____

(4.OA.B.4)

EXTRA PRACTICE

7. The number 60 is 12 times 5.

Write this comparison as a multiplication equation.

____

4.OA.B.4

8. Jane says that 9 is a factor of 99. Do you agree or disagree? Explain your reasoning.

____

____

____

4.OA.B.4

9. Julie has 4 piles of shirts in her drawer. Each pile has 6 shirts in it. How many shirts does Julie have? Explain your reasoning.

____

____

____

4.OA.A.2

10. There are 8 bike racks at the fair. Each bike rack has room for 12 bikes. How many bikes can fit on the bike racks? Explain your reasoning.

____

____

____

4.OA.A.2

EXTRA PRACTICE

11. Ms. Thompson needs 9 juice boxes for each lunch table. There are 9 lunch tables. How many juice boxes does Ms. Thompson need? What equation did you use?

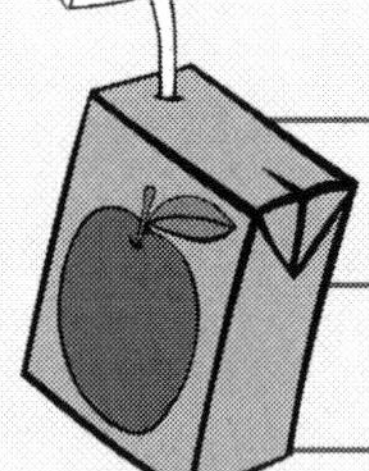

______

______

______

(4.OA.A.2)

12. Ms. Davis is getting pencils from the teacher work room. She needs 90 pencils. There are 18 pencils per box. How many boxes of pencils should Ms. Davis take from the work room?

Ms. Davis should take ________ boxes.

(4.OA.A.2)

13. What is the pattern rule in this sequence of numbers?

15, 20, 25

______

______

______

(4.OA.C.5)

14. What is the rule for this pattern?

3, 9, 27

A. Multiply by 3
B. Multiply by 2
C. Add 3
D. Add 6

(4.OA.C.5)

15. You are making a pattern that starts at 5 and has a rule to add 4. What is the 5th number in this pattern? Explain your reasoning.

____

____

____

____

4.OA.C.5

16. What is the pattern for the series of numbers? How do you know?

1, 10, 100, 1000?

____

____

____

____

4.OA.C.5

17. Daniel is thinking of a number that is 3 times the answer to 56 ÷ 8. What number is Daniel thinking of? How do you know?

____

____

____

____

4.OA.A.3

EXTRA PRACTICE

18. Araceli is trying to solve $49 \times 5 = x$. She said she is thinking the answer will be a little less than 250. Is this a reasonable estimate? Why or why not?

____

____

____

(4.OA.A.3)

19. Mark is picking apples at the orchard to make pies. He picked 35 apples from the first row of the orchard and 62 apples from the second row. He needs 7 apples for each pie he makes. How many pies will he be able to make with his apples?

A. 13 **B.** 14 **C.** 10 **D.** 15

(4.OA.A.3)

20. The cafeteria has 4 boxes of bananas with 30 bananas in each box. Seventy-two students took a banana each at lunch. How many bananas are left?

A. 48 **B.** 50 **C.** 102 **D.** 120

(4.OA.A.3)

NUMBER & OPERATIONS IN BASE TEN

PLACE VALUE - ROUNDING AND COMPARING 33

ADDITION AND SUBTRACTION 39

MULTIPLICATION AND DIVISION 46

CHAPTER REVIEW 51

EXTRA PRACTICE 58

1. Which statement correctly compares the value of the digit 4 in 243,188 to the value of the digit 4 in 947?

A. The digit 4 in 243,188 is 1,000 times larger than the digit 4 in 947.

B. The digit 4 in 947 is 10,000 times smaller than the digit 4 in 243,188.

C. The digit 4 in 243,188 is 10,000 larger than the digit 4 in 947.

D. The digit 4 in 947 is 10 times smaller than the digit 4 in 243,188.

4.NBT.A.1

2. Which statement correctly compares the value of the digit 6 in 63,143 to the value of the digit 6 in 4,006?

A. The digit 6 in 4,006 is one-thousandth the value of the digit 6 in 63,143.

B. The digit 6 in 4,006 is 100 times the the digit 6 in 63,143.

C. The digit 6 in 63,143 is 1,000 times the digit 6 in 4,006.

D. The digit 6 in 63,143 is 10,000 times the value of the digit 6 in 4,006.

4.NBT.A.1

3. There are 180 fourth-grade students at Hopewell Elementary School. The principal wants to place 20 students with 1 teacher in each fourth-grade class. Which expression can be used to determine the number of teachers needed?

A. $18 + 20$ **B.** 180×20 **C.** $18 \div 2$ **D.** $20 \div 180$

4.NBT.A.1

4. Mr. Walker has 350 pencils. He organizes the pencils in groups of 50. Which expression can be used to determine how many groups of 50 pencils Mr. Walker has?

A. $50 \div 350$ **B.** 350×50 **C.** $35 + 5$ **D.** $35 \div 5$

4.NBT.A.1

5. Evie reads at a speed of 140 words per minute. She is reading a short story that has 2,800 words. How many minutes will it take Evie to read the short story?

____

4.NBT.A.1

6. Fredo reads a book for 42 minutes. He can read at a speed of 100 words per minute. How many words are in the book?

____

4.NBT.A.1

7. Ian saved $37 to purchase a new tablet. The cost of the tablet is 10 times the amount he has saved. What is the cost of the tablet?

$_______________

4.NBT.A.1

8. Which number represents this expression written in expanded form?

789,000 – 10,899

A. 778,101

B. Seven hundred seventy-eight thousand, one hundred one

C. $(7 \times 100{,}000) + (7 \times 10{,}000) + (8 \times 1{,}000) + (1 \times 100) + (1 \times 1)$

D. $700{,}00 + 70{,}000 + 8{,}000 + 100 + 1$

4.NBT.A.2

9. Which number represents this expression written in expanded form?

$$1,000,000 - 4,009$$

A. $(1 \times 1,000,000) + (4 \times 1,000) + (9 \times 1)$

B. $(900,000 + 90,000 + 5,000) + 900 + 90 + 1$

C. $(1,000,000 + 4,000 + 9)$

D. $(900 + 95,000 + 900 + 90 + 1)$

4.NBT.A.2

10. Elise lives one-thousand, four hundred five miles from her grandparents. Her grandparents live seven hundred eighty miles from the Pacific Ocean. This tape diagram represents each location.

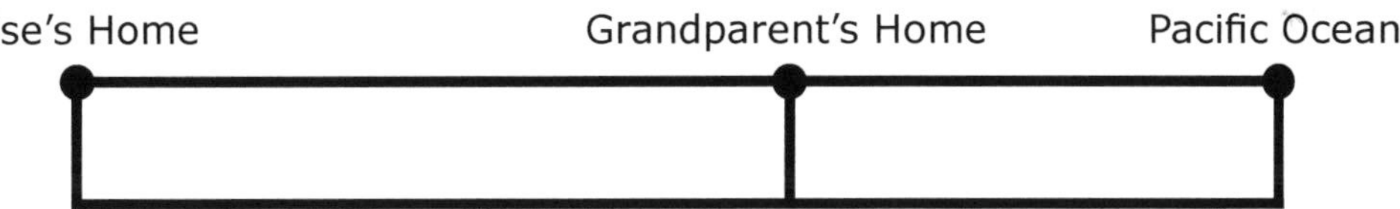

Which number represents the total distance, in miles, from Elise's home to the Pacific Ocean?

A. Two-thousand one hundred eighty-five

B. Six hundred twenty-five

C. Two-thousand two hundred thirty

D. One-thousand five hundred twenty-eight

4.NBT.A.2

PLACE VALUE – ROUNDING AND COMPARING

11. Mrs. Thomas wants to buy one of these cars.

The cost of both cars is $3,800 less than the original price. Write an inequality to compare the original costs of the cars.

____

(4.NBT.A.2)

12. Wes has two thousand eight hundred six dollars. Mary has two thousand eighty-six dollars. They decide to split the cost on a television which costs four hundred ninety dollars. How much more money will Wes have left than Mary?

____

(4.NBT.A.2)

13. Sean writes this number in expanded form.

150,008,740

100,000,000 + 50,000,000 + 8,000,000 + 700 + 40

Do you agree with his expression? Explain your reasoning.

____

____

____

(4.NBT.A.2)

14. Moises writes this number in expanded form.

58,006,040

50,000,000 + 8,000,000 + 6,000 + 400

Do you agree with his expression? Explain your reasoning.

____

____

____

(4.NBT.A.2)

15. Oscar travels miles to visit an amusement park. What is this number rounded to the nearest hundred?

A. 400 **B.** 500 **C.** 420 **D.** 430

(4.NBT.A.3)

16. This tape diagram shows the distance Ethan travels from his house to Dallas, Texas and then to Birmingham, Alabama.

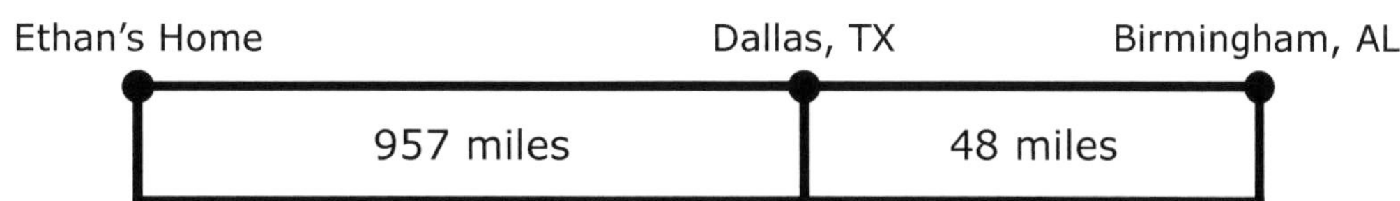

Which expression shows estimating the distance Ethan travels from his home to Birmingham, Alabama by rounding to the nearest hundred?

A. 900 + 700 **B.** 1,000 + 600
C. 1,000 + 700 **D.** 900 + 600

(4.NBT.A.3)

17. Bebe has 125 toys, 31 video games, and 28 stuffed animals. She receives 8 more toys, 3 more video games and 7 more stuffed animals on her birthday. Rounding to the nearest ten, write an equation to estimate the number of toys, video games, and stuffed animals Bebe has now.

____

(4.NBT.A.3)

18. Jacki has 134 pieces of candy. Jordan has 67 pieces of candy. Estimating to the nearest hundred, approximately how many pieces of candy do they have altogether?

____

(4.NBT.A.3)

19. Use the number line to explain why 67,400 rounded to the nearest ten-thousand is 70,000.

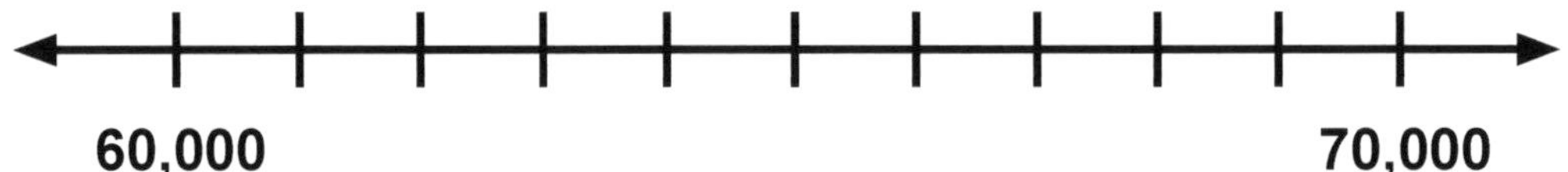

60,000 70,000

(4.NBT.A.3)

20. Jacob believes 18,999 rounded to the nearest thousand is 20,000. Irving believes 18,999 rounded to the nearest thousand is 19,000. Lorenzo believes 18,999 rounded to the nearest thousand is 18,000. Which person do you agree with? Explain your reasoning.

____

____

____

(4.NBT.A.3)

ADDITION AND SUBTRACTION

1. Which expression shows how these numbers should be added using the standard algorithm?

$$345 + 132$$

A. $(5+1)+(4+3)+(3+2)$

B. $(300+100)+(40+30)+(5+2)$

C. $(3+1)+(4+3)+(5+2)$

D. $(500+100)+(40+30)+(3+2)$

4.NBT.A.4

2. Which expression shows how these numbers should be added using the standard algorithm?

$$7{,}004 + 497$$

A. $(4+7)+(0+9)+(0+4)+7$

B. $(4+7)+(0+90)+(0+400)+7{,}000$

C. $(4+7)+(10+90)+(100+400)+7{,}000$

D. $(4+4)+(0+90)+(0+700)+7{,}000$

4.NBT.A.4

3. Which expression shows how these numbers should be subtracted using the standard algorithm?

$$4{,}065 - 2{,}948$$

A. $(15-8)+(50-40)+(100-900)+(3{,}000-2{,}000)$

B. $(8-5)+(60-40)+(900-0)+(4{,}000-2{,}000)$

C. $(8-5)+(6-4)+(9-0)+(4-2)$

D. $(15-8)+(50-40)+(1000-900)+(3{,}000-2{,}000)$

4.NBT.A.4

4. Sergio uses the standard algorithm to subtract these numbers.

$$\begin{array}{r} 402 \\ -\,178 \\ \hline \end{array}$$

Which statement could describe his first step?

A. Add 1 ten to the 8 ones.
B. Subtract 2 ones from the 8 ones.
C. Add 1 ten to the 2 ones.
D. Add 4 hundreds to the 2 ones.

4.NBT.A.4

5. Sergio uses the standard algorithm to subtract these numbers.

$$\begin{array}{r} 500 \\ -\,136 \\ \hline \end{array}$$

Which statement could describe his first step?

A. Add 1 ten to the 0 ones.
B. Subtract 0 ones from the 6 ones.
C. Add 5 hundreds to the 0 ones.
D. Subtract 0 from 36 ones.

4.NBT.A.4

6. Add using the standard algorithm.

$$2{,}475 + 5{,}967 = \underline{\hspace{3cm}}$$

A. 7,332 **B.** 7,343 **C.** 8,442 **D.** 8,441

4.NBT.A.4

7. Subtract using the standard algorithm.

$$5{,}095 - 497 = \_\_\_\_\_\_$$

A. 5.402 **B.** 4.598 **C.** 5.608 **D.** 4.301

4.NBT.A.4

8. Which expression can be added using these steps? ______

Step 1: Add the ones and regroup.

Step 2: Add the tens and regroup.

Step 3: Add the hundreds.

A	B	C
345	145	2,978
+ 289	+ 348	+ 1,218

4.NBT.A.4

9. Quintin is solving this equation using the standard algorithm.

$$4{,}056 + 2{,}280 = \_\_\_\_\_\_$$

Will he need to regroup? Explain your thinking.

____

____

____

4.NBT.A.4

ADDITION AND SUBTRACTION

10. Stan is solving this equation.

$$18{,}325 - 7{,}798 = \_\_\_\_\_\_\_\_$$

How many times will he need to regroup? Explain your reasoning.

______

______

______

______

4.NBT.A.4

11. Write the equation to represent the steps for subtracting with the standard algorithm.

Step 1: Regroup 1 ten as 10 ones. Subtract 9 ones from 17 ones.

Step 2: Regroup 1 hundred as 10 tens. Subtract 5 tens from 13 tens.

Step 3: Regroup 1 thousand as 10 hundreds. Subtract 6 hundreds from 12 hundreds.

Step 4: Subtract 3 thousands from 4 thousands.

______

______

______

______

4.NBT.A.4

12. Explain how to use the standard algorithm to subtract 176 from 374.

____

____

____

(4.NBT.A.4)

13. Which equation can be added using these steps? ________

Step 1: Add the ones.

Step 2: Add the tens and regroup.

Step 3: Add the hundreds and regroup.

Step 4: Add the thousands and regroup.

Equation A	Equation B	Equation C
4,345	5,191	2,978
+ 2,089	+ 7,948	+ 1,218

(4.NBT.A.4)

14. Which equation can be subtracted using these steps? ________

Step 1: Subtract the ones.

Step 2: Regroup the hundreds as tens then subtract.

Step 3: Regroups the thousands as hundreds then subtract.

Equation A	Equation B	Equation C
3,346	7,191	2,978
− 2,582	− 5,948	− 1,218

(4.NBT.A.4)

15. Which equation can be subtracted using these steps? ________

Step 1: Regroup the tens as ones then subtract.

Step 2: Regroup the hundreds as tens then subtract.

Step 3: Regroups the thousands as hundreds then subtract.

Equation A	Equation B	Equation C
4,892	7,981	2,000
− 1,531	− 5,548	− 1,123

(4.NBT.A.4)

16. Jenny is solving this equation using the standard algorithm.

$$7{,}000 - 3{,}158 = \_\_\_\_\_\_$$

Will she need to regroup? Explain your reasoning.

____

____

____

(4.NBT.A.4)

17. Casey uses the standard algorithm to add these numbers.

```
   1 1 1
   3̸,9̸1̸2
 + 1,098
 -------
   5,010
```

Do you agree with Casey's work? Explain your reasoning.

____

____

____

(4.NBT.A.4)

18. Explain how to use the standard algorithm to subtract 2,476 from 3,074.

____

____

____

4.NBT.A.4

19. Using the standard algorithm to add these numbers, which value would need to be regrouped?

$$\begin{array}{r} 3{,}415 \\ +\ 2{,}391 \\ \hline \end{array}$$

A. 6 Ones **B.** 10 Tens **C.** 7 Hundreds **D.** 5 Thousands

4.NBT.A.4

20. Write the equation to represent the steps for adding with the standard algorithm.

Step 1: Add 8 ones and 5 ones. Regroup 10 ones as 1 ten.

Step 2: Add 10 tens and 2 tens. Regroup 10 tens as 1 hundred.

Step 3: Add 4 hundreds and 5 hundreds.

Step 4: Add the 1 thousand and 2 thousands.

____

____

____

4.NBT.A.4

UNIT 3: MULTIPLICATION AND DIVISION

NAME: .... DATE: ....

NUMBER & OPERATIONS IN BASE TEN

1. Justin plays 45 basketball games each season. How many basketball games will he play in 12 seasons?

A. 57 **B.** 540 **C.** 460 **D.** 135

4.NBT.B.5

2. Mariam uses 27 beads for each necklace she creates. How many beads does she need for 18 necklaces?

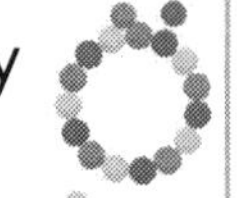

A. 45 **B.** 1,926 **C.** 243 **D.** 486

4.NBT.B.5

3. There are 28 students in each of the 11 fourth grade classes. Which equation could be used to find the total number of students in fourth grade?

A. $(30 \times 11) - 2 =$ ________ **B.** $(28 \times 10) + 1 =$ ________

C. $(30 \times 11) - 22 =$ ________ **D.** $(28 \times 20) - 9 =$ ________

4.NBT.B.5

4. A football team plays 15 games in a season. In each game, they score between 18 and 35 points.

Which number could represent the total number of points the team scores in a season?

A. 345 **B.** 250 **C.** 545 **D.** 195

4.NBT.B.5

5. Roseanne's birds eat between 12 and 18 ounces of food each day. Which number represents how many ounces of food the birds could eat in two weeks?

A. 154 **B.** 225 **C.** 260 **D.** 100

(4.NBT.B.5)

6. There are 2,150 students at Redwood Middle School. Half of the students buy lunch. Lunch costs $2. How much do the students at the school spend on lunch?

____

(4.NBT.B.5)

7. What is the value of this expression?

$$78 \times 49$$

____

(4.NBT.B.5)

8. There are 90 people going to a museum. One-third of the people are adults. The cost of an adult's ticket is $18, and the cost of child's ticket is $10. How much will this group of people spend on museum tickets?

____

(4.NBT.B.5)

9. Jamal uses this strategy to multiply and 2,348 and 8.

$$(2{,}000 \times 8) = 16{,}000$$
$$(300 \times 8) = 2{,}400$$
$$(40 \times 8) = 320$$
$$(8 \times 8) = 64$$

$$16{,}000 + 2{,}400 + 320 + 64 = 18{,}784$$

Do you agree with Jamal? Explain your reasoning.

____

____

____

4.NBT.B.5

10. Kara plants flowers in this flower bed.

11 ft.

4 ft.

If she plants 12 seeds in each square foot of the flower bed, how would you use multiplication to determine the number of seeds Kara plants?

____

4.NBT.B.5

11. A farmer picks 249 pears, and puts them in baskets. There are 12 pears in each basket. How many baskets does the farmer use?

A. 24 **B.** 21 **C.** 20 **D.** 12

4.NBT.B.6

12. Steve has 271 books to put on his bookshelves. Each shelf holds 21 books. How many shelves will he use for all the books?

A. 21 **B.** 12 **C.** 13 **D.** 14

4.NBT.B.6

13. The fourth-grade students have collected 348 pencils, 94 erasers, and 147 pens and put them together in small bags. Each bag must contain 8 pencils, 2 erasers, and 3 pens. How many complete bags will they create?

A. 49 **B.** 43 **C.** 47 **D.** 44

4.NBT.B.6

14. A grocery store has 546 cucumbers. There are 6 cucumbers in a package. Which strategy can be used to determine the total number of packages?

A. $(540 \div 6) + 6$ **B.** $(546 - 6) \div 6$
C. $(500 \div 6) + (40 \div 1)$ **D.** $(540 \div 6) + (6 \div 6)$

4.NBT.B.6

15. A grocery store has 378 potatoes delivered in crates. There are 9 potatoes in each crate.Which strategy could be used to determine the number of crates delivered?

A. $(360 \div 9) + 18$ **B.** $(360 \div 9) + (18 \div 9)$
C. $(300 \div 9) + (78 \div 1)$ **D.** $(378 - 18) \div 9$

4.NBT.B.6

16. Solve. $2,104 \div 7 =$ ________

A. 305 **B.** 300 R 4 **C.** 304 **D.** 30 R 4

4.NBT.B.6

MULTIPLICATION AND DIVISION

17. Two hundred nineteen people are volunteering at the homeless shelter. There are 9 people on each team of volunteers. How many complete teams are there?

____

(4.NBT.B.6)

18. An apartment building has 892 residents. There are 4 people living in each unit. How many units are in the apartment building?

____

(4.NBT.B.6)

19. Timothy's baseball team raises $12,267 selling towels and baseball bats. Each towel is sold for $6 and each baseball bat is sold for $9. They sell 1,482 towels. How many baseball bats does the team sell?

____

(4.NBT.B.6)

20. Mr. Welsh collects $1,290 from ticket sales for the school play. Each ticket costs $6. How would you determine the number of tickets sold?

____

____

____

(4.NBT.B.6)

1. Ada is calculating the value of this expression.

$$72{,}000 \div 800$$

Which statement describes the value of the expression?

A. The value of the expression is 100 times greater than $72 \div 8$.

B. The value of the expression is 10 times greater than $72 \div 8$.

C. The value of the expression is 100 times less than $72 \div 8$.

D. The value of the expression is 1,000 times less than $72 \div 8$.

(4.NBT.A.1)

2. Aki is calculating the value of this expression.

$$40 \times 10 \times 10$$

Which statement describes the value of the expression?

A. The value of the expression is 100 times greater than 4×1.

B. The value of the expression is 1,000 times greater than 4×1.

C. The value of the expression is 10 times greater than 4×1.

D. The value of the expression is 10,000 times greater than 4×1.

(4.NBT.A.1)

3. Jade's parents have $27,000 in a bank account. This amount is 30 times the amount they opened the account with. What is the difference, in dollars, between the amount of money Jade's parents saved and the amount of money they opened the account with?

____

(4.NBT.A.1)

CHAPTER REVIEW

4. Kari is comparing the digit 4 in these numbers.

15,450   48,796

> I think the 4 in 15,450 is 4 hundreds, and the 4 in 48,796 is 4 ten-thousands.
>
> Ten thousand is larger than one hundred.
>
> There are two place value positions between the hundreds place and the ten-thousands place.
>
> I think the 4 in 48,796 is 200 times larger than the 4 in 15,450.

Do you agree with Kari? Explain your reasoning.

____

____

____

(4.NBT.A.1)

5. There are seventeen thousand two people at the basketball game on Saturday. On Thursday, there are three thousand nineteen fewer people at the basketball game. How many people attend the basketball game on Thursday?

A. Fourteen thousand seventeen

B. Fourteen thousand ten

C. Thirteen thousand nine hundred eighty-three

D. Thirteen thousand nine hundred twelve

(4.NBT.A.2)

6. Lara has five hundred ninety dollars. Karla has four hundred seventy dollars. Lara earns seventy-five more dollars from babysitting. Karla earns twice the amount of money Lara earns from babysitting.

Write an inequality to compare the amounts of money Lara and Karla have after babysitting.

____

(4.NBT.A.2)

7. This table shows the populations of 5 different countries.

Country	Population
Austria	8,735,453
Sierra Leone	7,557,212
Gabon	2,025,137
Bahrain	1,492,584
Grenada	107,825

What is the difference, in written form, in population between the most populated country and the least populated country?

____

(4.NBT.A.2)

8. The population of El Salvador is 6,377,853. What is this number expressed in written form?

____

(4.NBT.A.2)

Ace Academic Publishing
ACHIEVING EXCELLENCE TOGETHER

CHAPTER REVIEW

9. Emil believes 23,021 rounded to the nearest hundred is 23,000.
Steven believes 23,021 rounded to the nearest hundred is 23,100.
Tomaso believes 23,021 rounded to the nearest thousand is 20,000.
Which person do you agree with? Explain your reasoning.

____

____

____

(4.NBT.A.3)

10. Larue wants to calculate the value of this expression:

$$5{,}927 + 2{,}725 + 13{,}078 + 495$$

He decides to round each number to the nearest thousand to estimate the value of the expression. Which number represents Larue's estimate using this strategy?

A. 22.000 **B.** 23.000 **C.** 20.000 **D.** 19.000

(4.NBT.A.3)

11. Use the number line to explain why 5,999 rounded to the nearest hundred is 6,000.

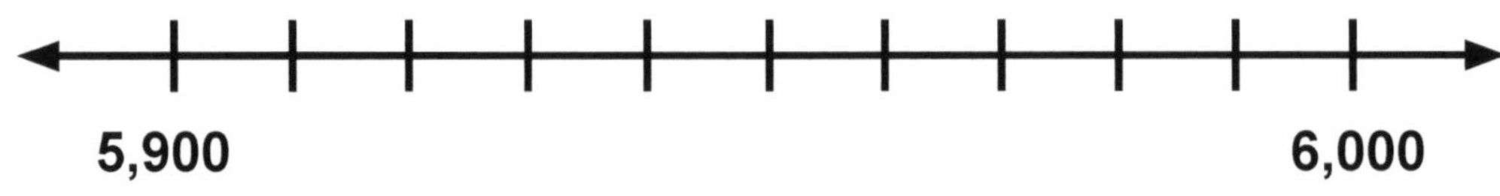

____

____

(4.NBT.A.3)

CHAPTER REVIEW

12. Using the standard algorithm to add these numbers, which values would need to be regrouped?

$$\begin{array}{r} 7{,}065 \\ +\ 1{,}816 \\ \hline \end{array}$$

A. 11 Ones **B.** 7 Tens **C.** 8 Hundreds **D.** 8 Thousands

4.NBT.B.4

13. Write out the steps you would use to add these numbers with the standard algorithm.

$$\begin{array}{r} 5{,}985 \\ +\ \ 709 \\ \hline \end{array}$$

____

____

____

4.NBT.B.4

14. Write an equation to represent the steps for adding with the standard algorithm.

Step 1: Add the 7 and 6 ones. Regroup 10 ones as 1 ten.

Step 2: Add the 8 tens and 4 tens. Regroup 10 tens as 1 hundred.

Step 3: Add 5 hundreds and 3 hundreds.

How do you know which equation to write?

____

____

4.NBT.B.4

CHAPTER REVIEW

15. What is the product of the multiplication expression represented by this area model?

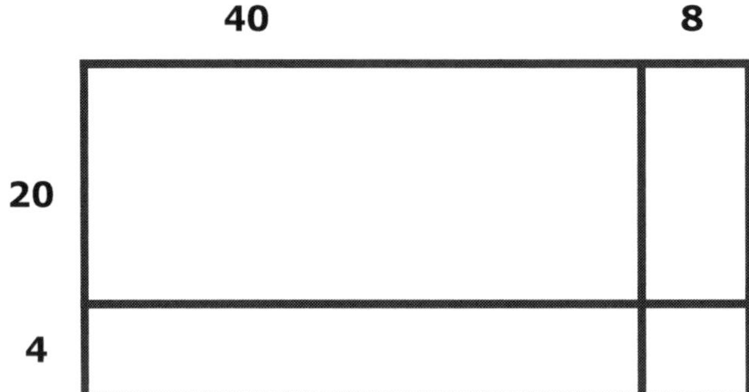

____

(4.NBT.B.5)

16. Dori has 25 rolls of quarters. Each roll of quarters contains 40 quarters. How many quarters does Dori have?

____

(4.NBT.B.5)

17. There are 31 students in Mr. Ngo's class and 27 students in Ms. Torres' class. Each student in Mr. Ngo's class pays $18 for a field trip to the zoo. Each student in Mr. Torres' class pays $20 for a field trip to the amusement park. Which class spends more money for their field trip?

____

____

____

(4.NBT.B.5)

18. A football stadium has 6 sections, and the same number of seats in each section. There were 2,742 attendees at the last football game. If the attendees are assigned tickets evenly between the sections, how many people are in each section?

____

4.NBT.B.6

19. Solve.

$$928 \div 6$$

A. 154 R 4 **B.** 158 **C.** 15 R 4 **D.** 155

4.NBT.B.6

20. Jamal uses this strategy to divide these numbers.

$$6{,}954 \div 3$$

$$(6{,}000 \div 3) = 2{,}000$$
$$(900 \div 3) = 300$$
$$(54 \div 3) = 18$$

$$2{,}000 + 300 + 18 = 2{,}318$$

Do you agree with Jamal? Explain your reasoning.

____

____

____

4.NBT.B.6

EXTRA PRACTICE

NAME: .... DATE: ....

NUMBER & OPERATIONS IN BASE TEN

EXTRA PRACTICE

1. The population of Pine City is one-tenth the population of Rush City. The population of Rush City is one-tenth the population of Mora. The population of Mora is 80,000 people.

What is the population of Pine City?

____

4.NBT.A.1

2. Which numbers are represented by each point?

- Point K is 100 times Point M.
- Point K is 100 less than Point L.
- Point L is 10 times Point N.

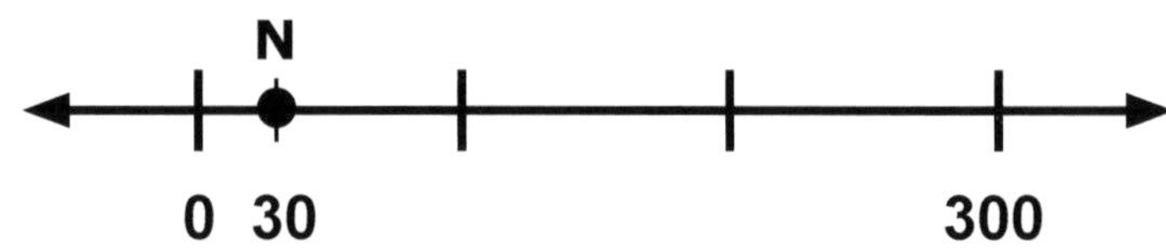

Explain your reasoning.

____

____

____

4.NBT.A.1

3. A toy store sells a board game for $11. The store makes $5,500 selling the board games. How many board games does the store sell?

A. 5,000 **B.** 500 **C.** 5 **D.** 5,489

4.NBT.A.1

4. This table shows the number of bracelets Jenna makes each month.

Month	Number of Bracelets
January	18
February	32
March	17
April	63
May	40

Between June and December, Jenna makes 10 times the number of bracelets she made between January and May. How many bracelets does Jenna make between January and December?

____

4.NBT.A.1

5. Mr. Williams wants to buy one of these two cars.

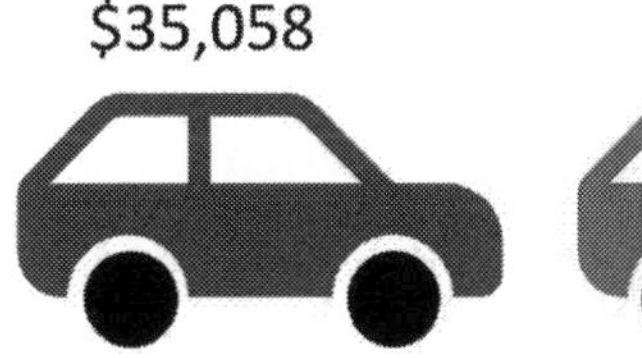

He has $40,000 to use to purchase a car. Which inequality correctly compares the amount of money Mr. Williams would have left after purchasing either car?

A. 35,508 > 35,508 **B.** 35,508 < 35,058

C. 4,492 > 4,942 **D.** 4,492 < 4,942

4.NBT.A.2

EXTRA PRACTICE

6. Write a number using these clues.

- The number has 6 digits.
- The value of the digit 2 is.
- The value of the digit 6 is.
- The value of the digit in the ones place is half the value of the digit in the thousands place.
- Each other digit is 1 more than the digit to the right.

____

(4.NBT.A.2)

7. How does the digit 3 in the number 123,456 compare to the digit 3 in 457,387?

____

____

____

(4.NBT.A.2)

8. Lucy is comparing these two numbers.

$$(8 \times 100{,}000) + (4 \times 1{,}000) + (5 \times 100) + (2 \times 10)$$

$$(1 \times 1{,}000{,}000) + (4 \times 100{,}000) + (3 \times 10{,}000) + (2 \times 1{,}000) + (1 \times 1)$$

What inequality compares these numbers? Explain your reasoning.

____

____

____

(4.NBT.A.2)

9. Terrell has a coin collection with three different sets of coins.

- The first set of coins is valued at $3,687.
- The second set of coins is valued at $1,208.
- The third set of coins is valued at $978.
- If he rounds each value to the nearest hundred, what is the approximate value of Terrell's coin collection?

A. $5.900 **B.** $5.000 **C.** $5.700 **D.** $6.000

4.NBT.A.3

10. This bar graph shows the pieces of candy collected by 4 friends.

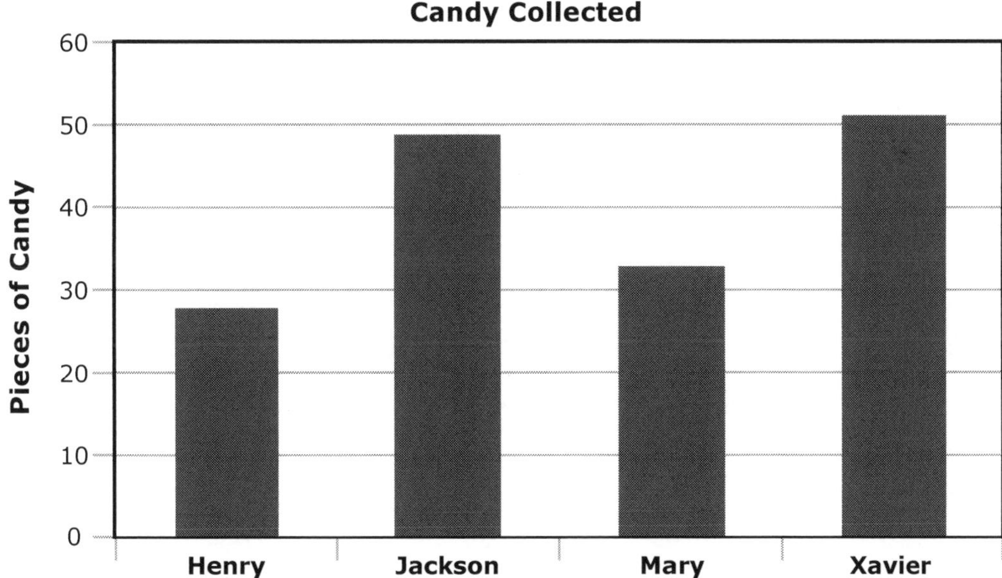

If each person's collection is rounded to the nearest ten, approximately how many pieces of candy do they have altogether?

____

4.NBT.A.3

EXTRA PRACTICE

11. Use the clues to determine this number.

- A 5-digit number rounds to 35,000 when rounded to the nearest thousand.
- The sum of the digits in the ten thousand and thousand places is equal to the digit in the ones place.
- The digit in the ones place is 7.
- The digit in the tens place is 2.
- The digit in the hundreds place is 1 more than the digit in the ones place.

____

4.NBT.A.3

12. Evelyn uses the standard algorithm to subtract these numbers.

$$\begin{array}{r} 2{,}000 \\ -\ 1{,}123 \\ \hline 1{,}123 \end{array}$$

Do you agree with her strategy? Explain your reasoning.

____

____

____

____

____

4.NBT.B.4

13. Maira is solving this equation using the standard algorithm.

$$453 - 284 = \_\_\_\_\_\_$$

Will she need to regroup? Explain your reasoning.

____

____

____

(4.NBT.B.4)

14. Ben uses the standard algorithm to subtract these numbers.

```
     11 13
   0 1̸ 3̸ 17
   1̸2,4̸7̸8
 −  8,497
 ---------
    3,981
```

Do you agree with Ben? Explain your reasoning.

____

____

____

(4.NBT.B.4)

15. There are 3,129 people at the State Fair on Monday. Twice as many people attend the State Fair on Tuesday. The number of people attending the State Fair on Wednesday, Thursday, and Friday is the same, each day, as the number of people who attend on Tuesday. Which expression could be used to find the number of people who attend the State Fair all 5 days?

A. $5 \times 3{,}129 \times 2$

B. $3{,}129 + (3{,}129 \times 2)$

C. $(3{,}129 \times 2) + (3{,}129 \times 2) + (3{,}129 \times 2) + (3{,}129 \times 2)$

D. $3{,}129 + 4(3{,}129 \times 2)$

(4.NBT.B.5)

Ace Academic Publishing
ACHIEVING EXCELLENCE TOGETHER

NUMBER & OPERATIONS IN BASE TEN

EXTRA PRACTICE

16. What is the product of the multiplication expression represented by this area model?

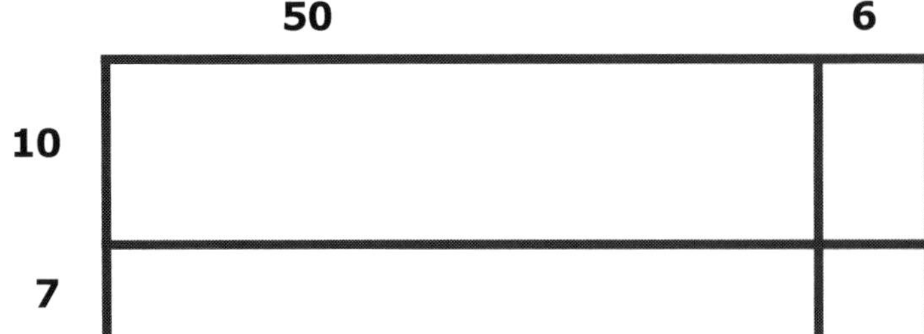

______

(4.NBT.B.5)

17. Explain why both strategies can be used to find the product of 88 and 12.

Strategy A: $(88 \times 12) = (176 \times 6) = (100 \times 6) + (70 \times 6) + (6 \times 6) = 1{,}056$

Strategy B: $(80 \times 12) + (8 \times 12) = 960 + 96 = 1{,}056$

______

______

(4.NBT.B.5)

18. There are 324 people auditioning for a music competition. After the first round, half of the people are eliminated. The remaining competitors are placed on 9 teams. How would you calculate how many people are on each team?

______

______

(4.NBT.B.6)

19. Timothy's baseball team raises $2,226 selling 303 towels and baseball bats, combined.

- Each towel is sold for $7 and each baseball bat is sold for $8.
- They earn $840 from the baseball bats.
- How much more does the team earn from selling towels than baseball bats?

4.NBT.B.6

20. The fourth-grade students collected 275 pencils, 180 erasers, and 328 pens. They put the pencils, erasers, and pens together in small bags. Each bag had 10 pencils, 6 erasers, and 9 pens. How many complete bags did the students create?

4.NBT.B.6

NUMBER & OPERATIONS FRACTIONS

EQUIVALENT FRACTIONS AND ORDERING FRACTIONS 69

FRACTION OPERATIONS 75

DECIMAL CONVERSION AND COMPARISON 82

CHAPTER REVIEW 89

EXTRA PRACTICE 96

1. The shaded part of the model below represents a fraction of the total area. The entire bar model represents 1.

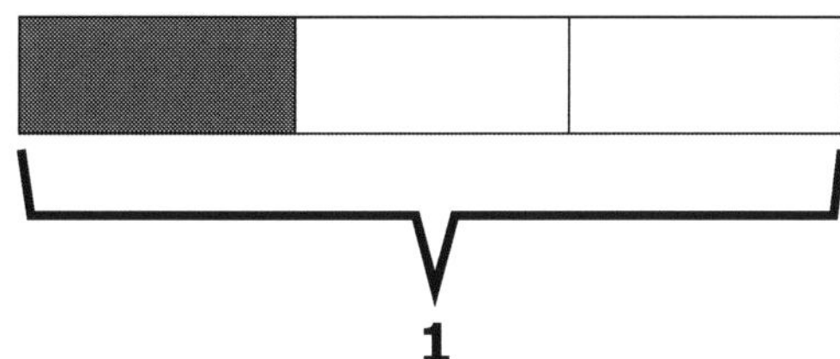

Which model below represents an equivalent fraction to the bar model above?

A. 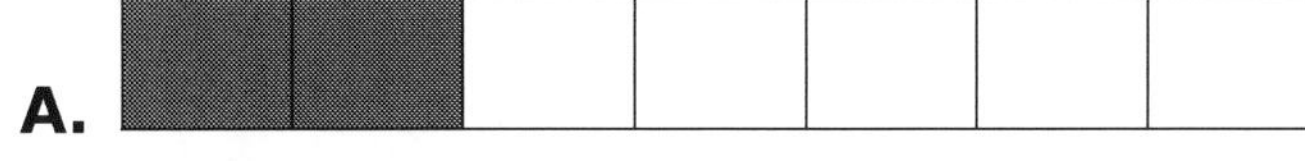

B. 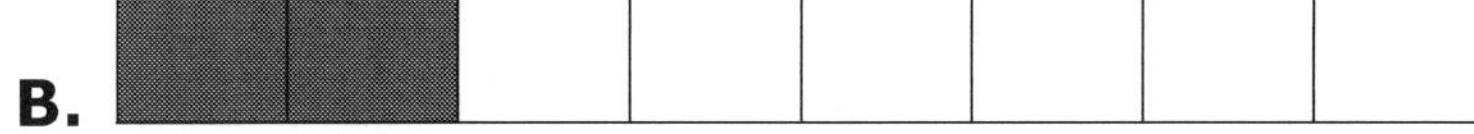

C. 

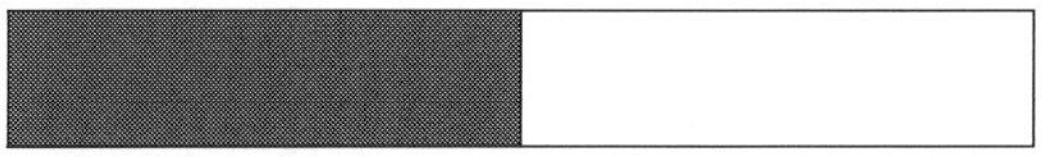

D.

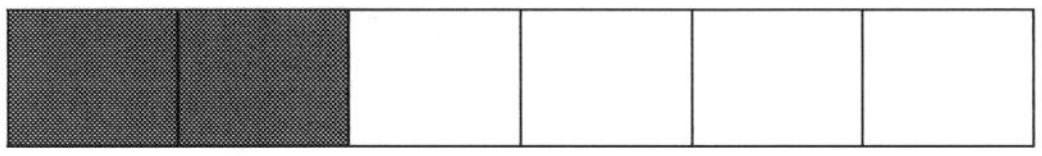

4.NF.A.1

2. Which fraction is equivalent to $\frac{4}{10}$?

A. $\frac{6}{12}$ **B.** $\frac{2}{6}$ **C.** $\frac{40}{100}$ **D.** $\frac{3}{9}$

4.NF.A.1

3. Which fraction is equivalent to the ratio of the shaded region and the total area of this model?

A. $\frac{1}{4}$ **B.** $\frac{2}{3}$

C. $\frac{1}{2}$ **D.** $\frac{2}{2}$

(4.NF.A.1)

4. What fraction is equivalent to the part of the bar model that is NOT shaded?

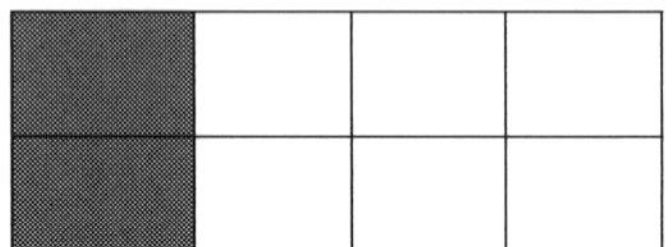

A. $\frac{1}{4}$ **B.** $\frac{3}{4}$

C. $\frac{1}{2}$ **D.** $\frac{2}{2}$

(4.NF.A.1)

5. Mia has set of 10 plastic stars. She has 6 blue stickers and 4 red stickers. Which fraction is equivalent to the fraction ratio of the red stars to total stars?

A. $\frac{2}{5}$ **B.** $\frac{4}{8}$ **C.** $\frac{1}{2}$ **D.** $\frac{5}{10}$

(4.NF.A.1)

6. The point on the number line is located on a fraction.

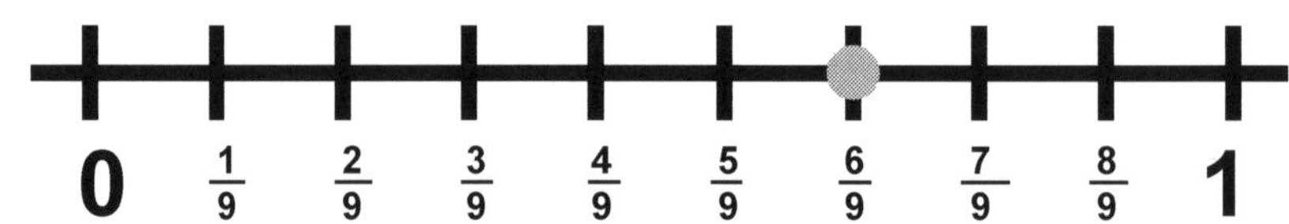

Which fraction could be represented by the point on the number line?

A. $\frac{4}{7}$ **B.** $\frac{3}{6}$ **C.** $\frac{2}{3}$ **D.** $\frac{1}{6}$

(4.NF.A.1)

7. Use bar models, shapes, circle graphs, or number lines to show 2 equivalent fractions for $\frac{1}{5}$.

(4.NF.A.1)

8. The shaded part of the model below represents a fraction of the total area of the model.

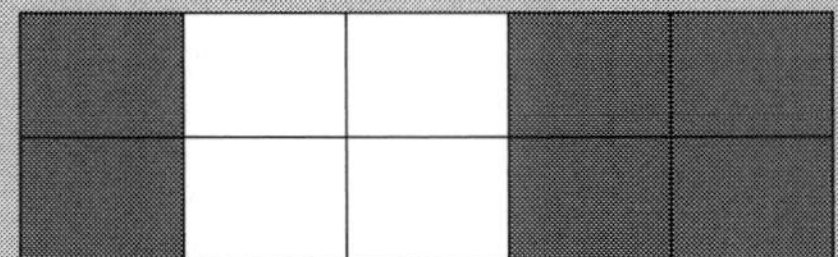

Which of the following fractions are equivalent to the ratio of the unshaded portion to the shaded portion in the model?

$\frac{12}{20}$ $\frac{3}{5}$ $\frac{8}{18}$ $\frac{2}{5}$ $\frac{18}{30}$

(4.NF.A.1)

Ace Academic Publishing
ACHIEVING EXCELLENCE TOGETHER

9. Harry has 18 chocolates. Of the 18 chocolates, 4 of them are white chocolate. The rest are milk chocolate. Which fraction is equivalent to the fraction of chocolates that are milk chocolates?

A. $\frac{4}{6}$ **B.** $\frac{7}{9}$ **C.** $\frac{12}{16}$ **D.** $\frac{5}{7}$

(4.NF.A.1)

10. Which fraction is equivalent to $\frac{8}{12}$? Use the model below to help you solve.

A. $\frac{4}{8}$ **B.** $\frac{2}{3}$

C. $\frac{6}{9}$ **D.** $\frac{3}{4}$

(4.NF.A.1)

11. Which fraction has the smallest value?

A. $\frac{3}{10}$ **B.** $\frac{2}{3}$ **C.** $\frac{1}{5}$ **D.** $\frac{3}{4}$

(4.NF.A.2)

12. Which fraction has the greatest value?

A. $\frac{5}{8}$ **B.** $\frac{1}{5}$ **C.** $\frac{2}{4}$ **D.** $\frac{11}{12}$

(4.NF.A.2)

13. The shaded model below shows two fractions.

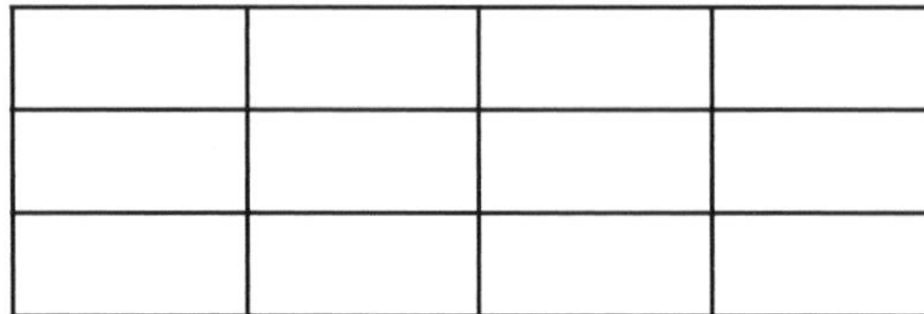

Which expression correctly compares the shaded part of the fraction models?

A. $\frac{1}{3} > \frac{4}{6}$ **B.** $\frac{4}{6} > \frac{1}{3}$ **C.** $\frac{2}{6} > \frac{1}{3}$ **D.** $\frac{2}{6} > \frac{2}{3}$

(4.NF.A.2)

14. The shaded models below show two fractions.

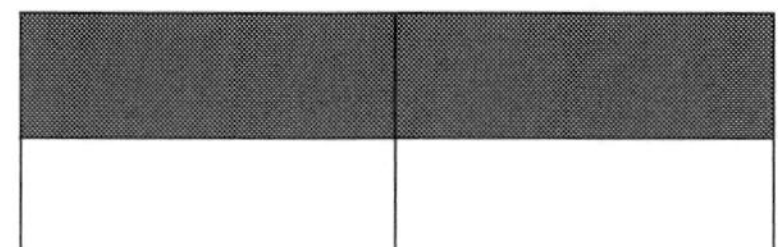

Which expression correctly compares the shaded fraction models?

A. $\frac{2}{4} > \frac{1}{2}$ **B.** $\frac{2}{4} = \frac{1}{2}$ **C.** $\frac{2}{4} < \frac{1}{2}$ **D.** $\frac{1}{4} > \frac{1}{2}$

(4.NF.A.2)

15. Which expression correctly compares two fractions?

A. $\frac{1}{4} < \frac{3}{6}$ **B.** $\frac{1}{6} > \frac{3}{4}$ **C.** $\frac{2}{8} > \frac{7}{9}$ **D.** $\frac{8}{12} < \frac{4}{10}$

(4.NF.A.2)

16. Choose the fraction that makes the comparison statement true.

$$\frac{2}{4} < \_\_\_\_\_\_\_$$

A. $\frac{2}{8}$ **B.** $\frac{4}{12}$ **C.** $\frac{7}{10}$ **D.** $\frac{3}{8}$

(4.NF.A.2)

17. Order the fractions from least to greatest.

$$\frac{60}{100} \quad \frac{2}{5} \quad \frac{20}{25} \quad \frac{4}{20}$$

Least Greatest

______ ______ ______ ______

(4.NF.A.2)

18. Compare the fractions below to $\frac{1}{2}$.

$\frac{4}{5}$ $\frac{6}{9}$ $\frac{7}{10}$ $\frac{3}{7}$ $\frac{7}{14}$ $\frac{50}{100}$ $\frac{3}{6}$

Which of the fractions are greater than $\frac{1}{2}$?
Which of the fractions are equivalent to $\frac{1}{2}$?

____

____

(4.NF.A.2)

19. Jonah studied for $\frac{3}{4}$ of an hour for his math test. Gino studied for $\frac{6}{12}$ of an hour for his spelling test. Kara studied for $\frac{2}{6}$ of an hour for her reading test. Ashley studied for $\frac{7}{10}$ of an hour for her science test. What was the largest amount of time, in hours, studied?

A. $\frac{7}{10}$ **B.** $\frac{6}{12}$ **C.** $\frac{3}{4}$ **D.** $\frac{2}{6}$

(4.NF.A.2)

20. Sam ate $\frac{6}{8}$ of his sandwich. Emma ate $\frac{4}{5}$ of her sandwich, and Kevin ate $\frac{2}{4}$ of his sandwich. Whose fraction has a value that is in between the values of the other two fractions?

A. Sam's
B. Sam's and Emma's
C. Kevin's
D. Emma's

(4.NF.A.2)

UNIT 2: FRACTION OPERATIONS

NUMBER & OPERATIONS – FRACTIONS

1. Which addition expression represents the total shaded parts in this fraction model?

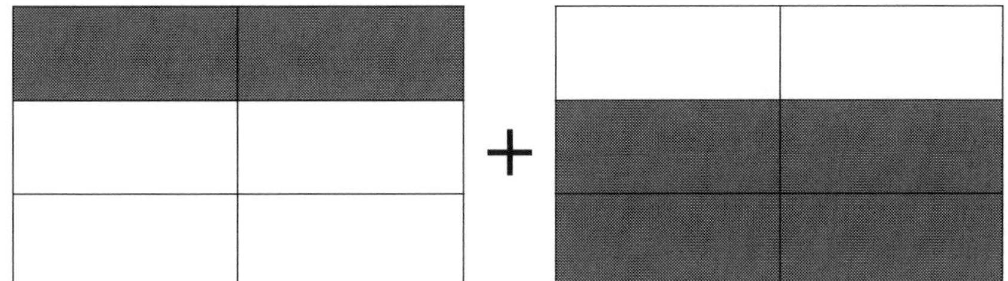

A. $\frac{2}{3}+\frac{1}{2}$ B. $\frac{2}{6}+\frac{3}{6}$ C. $\frac{2}{6}+\frac{4}{6}$ D. $\frac{4}{6}+\frac{3}{6}$

4.NF.B.3

2. Which expression has a sum that equals $\frac{4}{5}$?

A. $\frac{1}{4}+\frac{1}{4}+\frac{1}{4}+\frac{1}{4}$ B. $\frac{1}{5}+\frac{1}{5}+\frac{1}{5}+\frac{1}{5}$

C. $\frac{1}{5}+\frac{1}{5}+\frac{1}{5}+\frac{1}{5}+\frac{1}{5}$ D. $\frac{1}{4}+\frac{1}{4}+\frac{1}{4}$

4.NF.B.3

3. Find the sum: $\frac{2}{8}+\frac{5}{8}$

A. $\frac{7}{8}$ B. $\frac{7}{16}$ C. $\frac{6}{8}$ D. $\frac{10}{64}$

4.NF.B.3

4. Find the difference: $\frac{9}{10}-\frac{5}{10}$

A. $\frac{2}{10}$ B. $\frac{3}{5}$ C. $\frac{3}{10}$ D. $\frac{4}{10}$

4.NF.B.3

FRACTION OPERATIONS

5. Which expression represents the shaded portion of the figure below?

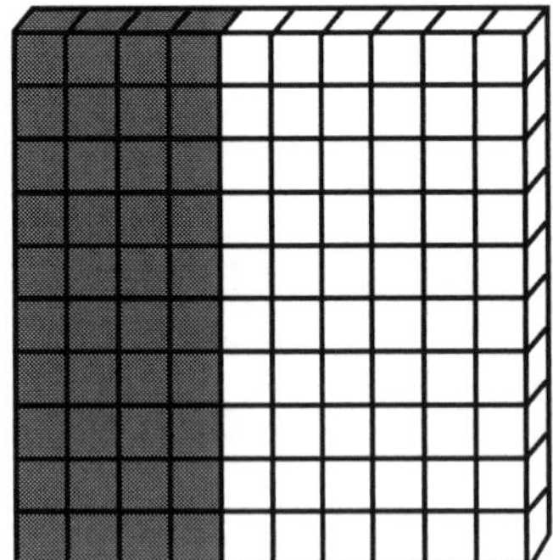

A. $\frac{1}{10} + \frac{1}{10} + \frac{1}{10}$

B. $\frac{4}{10} + \frac{4}{10} + \frac{4}{10}$

C. $\frac{1}{10} + \frac{1}{10} + \frac{1}{10} + \frac{1}{10}$

D. $\frac{1}{4} + \frac{1}{4} + \frac{1}{4} + \frac{1}{4}$

(4.NF.B.3)

6. Which sum is greater than 1?

A. $\frac{4}{8} + \frac{3}{8}$ **B.** $\frac{2}{5} + \frac{2}{5}$ **C.** $\frac{9}{18} + \frac{8}{18}$ **D.** $\frac{7}{12} + \frac{6}{12}$

(4.NF.B.3)

7. Dane ate $\frac{4}{8}$ of a whole pizza. How much of the pizza is left?

A. $\frac{5}{8}$ **B.** $\frac{4}{8}$ **C.** $\frac{3}{8}$ **D.** $\frac{2}{8}$

(4.NF.B.3)

8. Choose the expression that shows adding $3\frac{3}{4} + 2\frac{2}{4}$ using equivalent fractions.

A. $\frac{15}{4} + \frac{10}{4}$ **B.** $\frac{6}{4} + \frac{4}{4}$ **C.** $\frac{13}{4} + \frac{8}{4}$ **D.** $\frac{12}{4} + \frac{10}{4}$

(4.NF.B.3)

9. This week, Nancy spent $3\frac{2}{5}$ hours reading on Friday and $2\frac{4}{5}$ hours reading on Saturday. How many groups of $\frac{1}{5}$ hours did Nancy read this week?

A. 6 **B.** 5 **C.** 11 **D.** 31

(4.NF.B.3)

10. Karl must practice the violin for $\frac{1}{4}$ hours each day. He has practiced $7\frac{3}{4}$ hours. How many days has Karl practiced the violin?

A. 7 days **B.** 28 days **C.** 31 days **D.** 70 days

(4.NF.B.3)

11. Fifteen people attended a party. Two-fifths of the people brought a gift. Which model represents the portion of the people who did not bring a gift?

A. 

B. 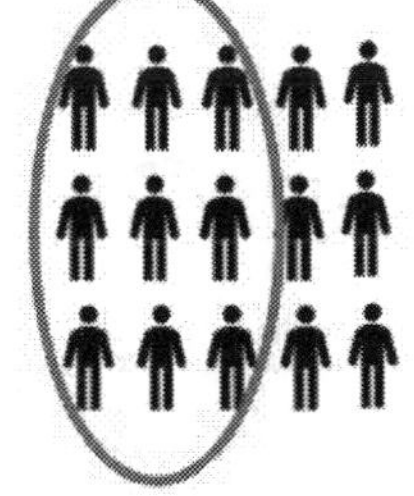

C. 

D. 

(4.NF.B.4)

12. Three-fourths of the students in Alex's class are going to the fair. One-third of those students are riding on a bus to get there. Which model shows the number of students riding a bus going to the fair?

A. 

B. 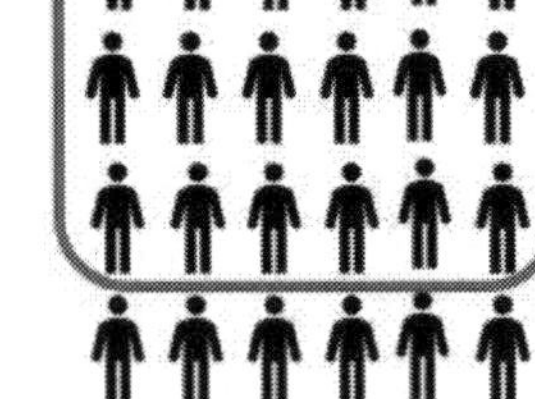

C. 

D. 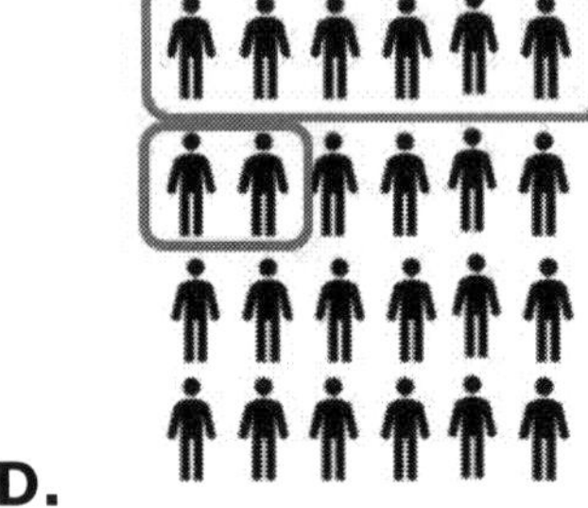

(4.NF.B.4)

13. Chloe is wrapping gifts. She uses this length of ribbon to wrap 1 gift.

How much ribbon, in feet, does Chloe need to wrap 5 gifts?

____

(4.NF.B.4)

14. Chase has green, red, and yellow pencils.

- One-eighth of the pencils are green.
- Three-eighths of the pencils are red.
- There are 32 pencils.

Which equation can be used to determine the number of yellow pencils?

A. $\frac{4}{8}(32)=16$ **B.** $\frac{1}{8}(32)=4$ **C.** $\frac{3}{8}(32)=12$ **D.** $\frac{4}{16}(32)=8$

(4.NF.B.4)

15. Sanjay fills this measuring cup with peanuts.

Peanuts

He adds the same amount of peanuts to 7 more measuring cups. Which expression can be used to find the total amount peanuts Sanjay has?

A. $\frac{5}{8}+8$ **B.** $\frac{5}{8}\times 8$ **C.** $\frac{5}{8}\times 7$ **D.** $\frac{5}{8}+7$

(4.NF.B.4)

16. Ronnell reads $\frac{1}{3}$ of a book in 5 days. He reads the same number of pages each day. How many days will it take Ronnell to finish the book? Explain how to find the answer.

____

____

____

(4.NF.B.4)

17. Walter uses the water and lemon juice, in the two measuring cups below, to make 1 serving of lemonade.

How much water and lemon juice, combined, will he need to make 5 servings of lemonade?

A. $\frac{15}{3}$ cups **B.** $6\frac{2}{3}$ cups

C. $\frac{5}{3}$ cups **D.** $3\frac{1}{3}$ cups

(4.NF.B.4)

18. Henrietta bought 18 bananas. She used $\frac{2}{3}$ of the bananas for a recipe. How many bananas did she use for the recipe?

A. 13 **B.** 6 **C.** 16 **D.** 12

(4.NF.B.4)

19. Fill in the blanks to write an expression matching this model.

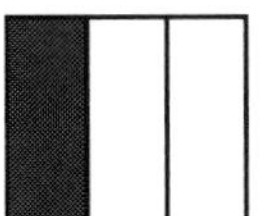
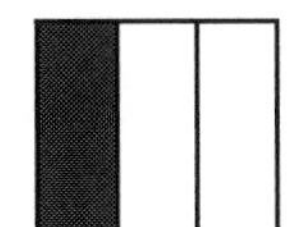
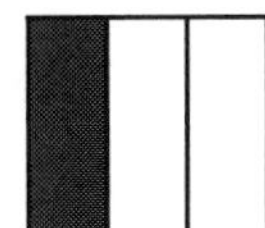
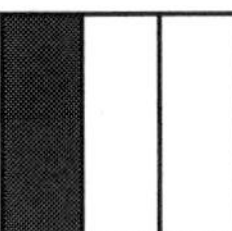

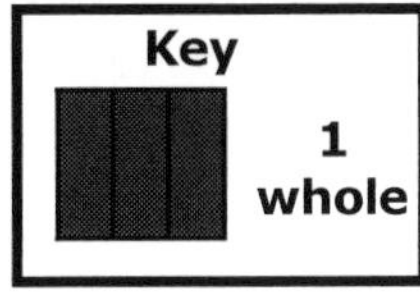

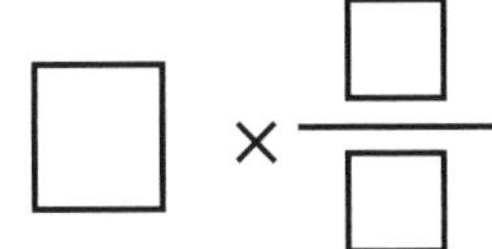

(4.NF.B.4)

20. A recipe requires $\frac{5}{8}$ cups of fruit and $\frac{3}{8}$ cups of juice to make 2 servings. Sebastian wants to make 14 servings. How much fruit and how much juice will he need?

(4.NF.B.4)

UNIT 3: DECIMAL CONVERSION AND COMPARISON

NAME: .... DATE: ....

NUMBER & OPERATIONS – FRACTIONS

DECIMAL CONVERSION AND COMPARISON

1. Erin measures the length of this leaf.

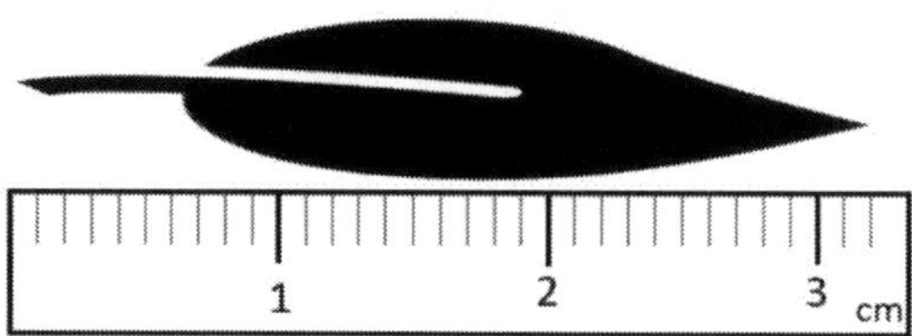

Which fraction is equivalent to the length of the leaf?

A. $3\frac{2}{100}$ cm **B.** $3\frac{20}{100}$ cm **C.** $3\frac{2}{9}$ cm **D.** $3\frac{10}{2}$ cm

4.NF.C.5

2. Which fraction does Point A on this number line represent?

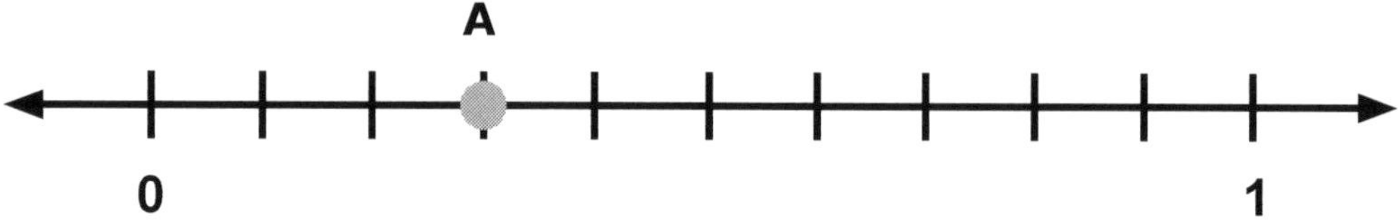

A. $\frac{4}{10}$ **B.** $\frac{5}{10}$ **C.** $\frac{30}{100}$ **D.** $\frac{50}{100}$

4.NF.C.5

3. What fraction represents the sum of Model A and Model B?

Model A **Model B**

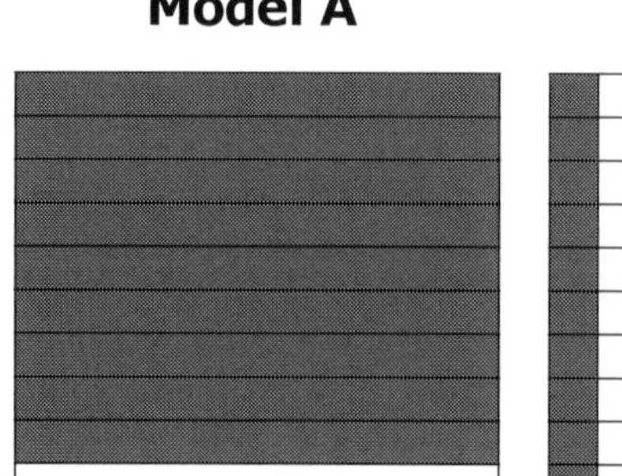

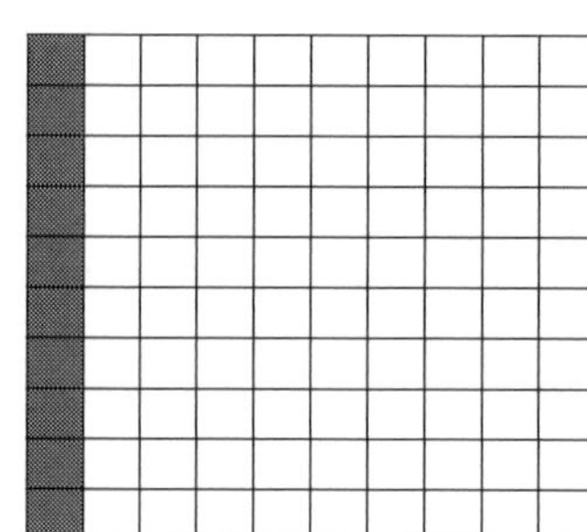

____

4.NF.C.5

4. William determines the distance between Point **X** and Point **Y** on this number line is $\frac{4}{10}$.

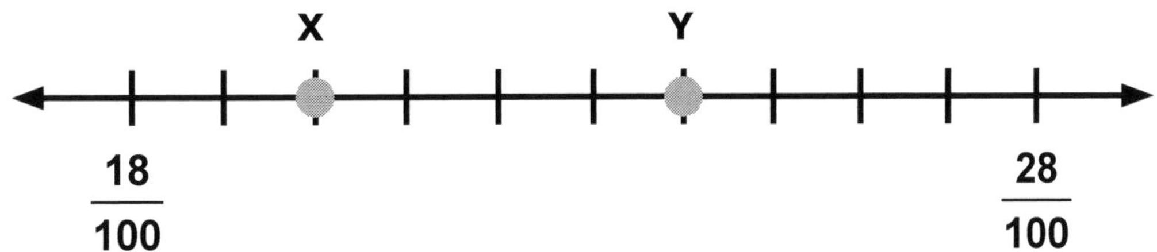

Do you agree with William? Explain your reasoning.

____

____

____

4.NF.C.5

5. What fraction represents the sum of Model A and Model B?

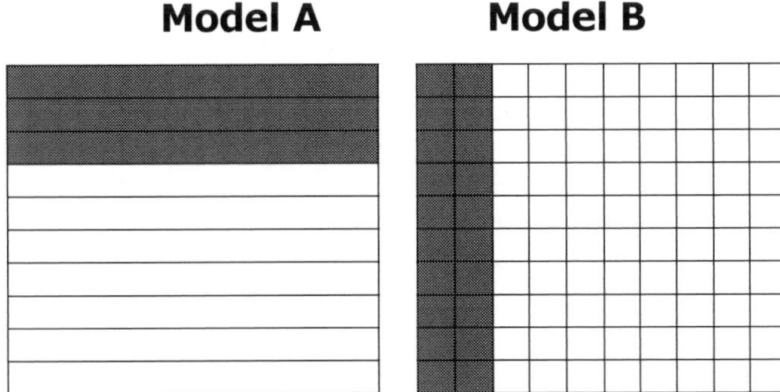

____

4.NF.C.5

DECIMAL CONVERSION AND COMPARISON

6. Which statement correctly compares the values represented by shaded part of these models?

Model A **Model B**

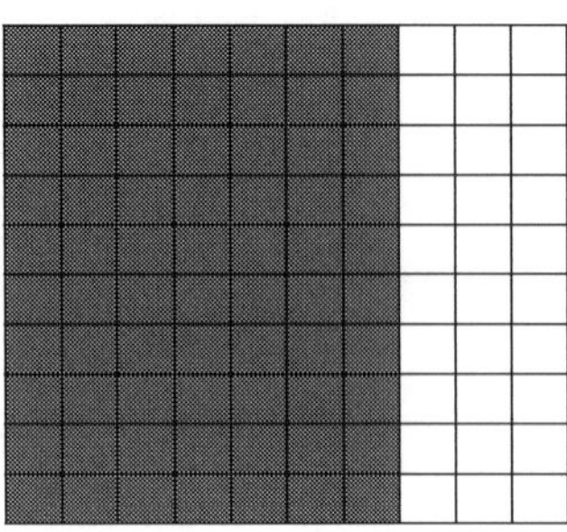

A. $\frac{7}{10} = \frac{70}{100}$ **B.** $\frac{7}{3} = \frac{70}{30}$ **C.** $\frac{7}{10} = \frac{70}{10}$ **D.** $\frac{3}{10} = \frac{30}{100}$

(4.NF.C.5)

7. Mara has a pot with a capacity of $200\frac{5}{10}$ ounces.

Which expression can she use to write this fraction with 100 as a denominator?

A. $200\ \frac{5+90}{10+90} = 200\frac{95}{100}$ **B.** $200\ \frac{5\times 10}{10\times 10} = 200\frac{50}{100}$

C. $200\ \frac{5}{10\times 10} = 200\frac{5}{100}$ **D.** $200\times 10\ \frac{5\times 10}{10\times 10} = 2{,}000\frac{50}{100}$

(4.NF.C.6)

8. The length of a pen is $12\frac{4}{10}$ centimeters. Which decimal is equivalent to the length of the pen?

A. 12.04 cm **B.** 12.4 cm **C.** 124.0 cm **D.** 12.410 cm

(4.NF.C.6)

9. An eraser is $6\frac{3}{10}$ centimeters long. Which equation about the length of this eraser is true?

A. $0.63 = 6\frac{3}{10}$ **B.** $0.063 = 6\frac{3}{10}$ **C.** $6.03 = 6\frac{3}{10}$ **D.** $6.3 = 6\frac{3}{10}$

(4.NF.C.6)

10. How could the length of these scissors be represented in decimal form?

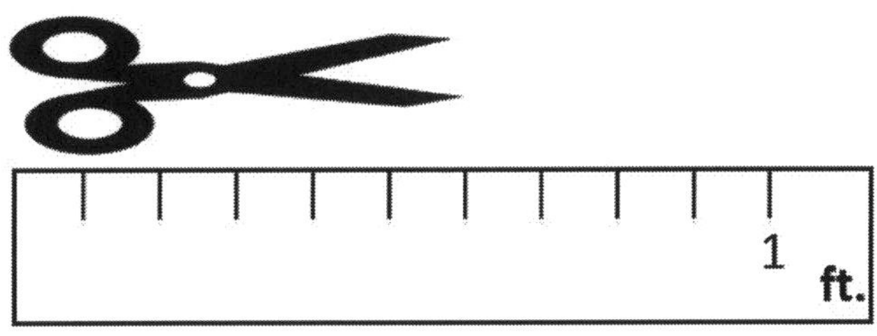

A. Divide 6 by 100 **B.** Multiply 6 by 100
C. Divide 6 by 10 **D.** Multiply 6 by 100

(4.NF.C.6)

11. Terrence says the length of his math book is $32\frac{74}{100}$ centimeters. Which decimal shows another way to express this length?

A. 32.74100 **B.** 32.074 **C.** 3274.100 **D.** 32.74

(4.NF.C.6)

12. A group of 100 students were asked to identify their favorite color. This table shows how many of the students li ked each of the colors.

Fraction	red	blue	yellow	green	pink
Color	$\frac{24}{100}$	$\frac{10}{100}$	?	$\frac{1}{100}$	$\frac{46}{100}$

Which decimal represents the fraction of students who like yellow?

____

(4.NF.C.6)

Ace Academic Publishing
ACHIEVING EXCELLENCE TOGETHER

13. Pria determines the distance between Point A and Point C on this number line is 0.2 units.

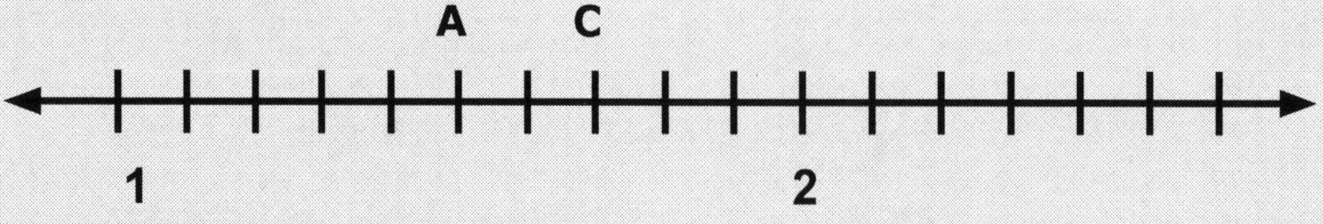

Do you agree with Pria? Explain your reasoning.

____

____

____

4.NF.C.6

14. Eva uses this model to represent a decimal.

Parth uses this model to represent a decimal.

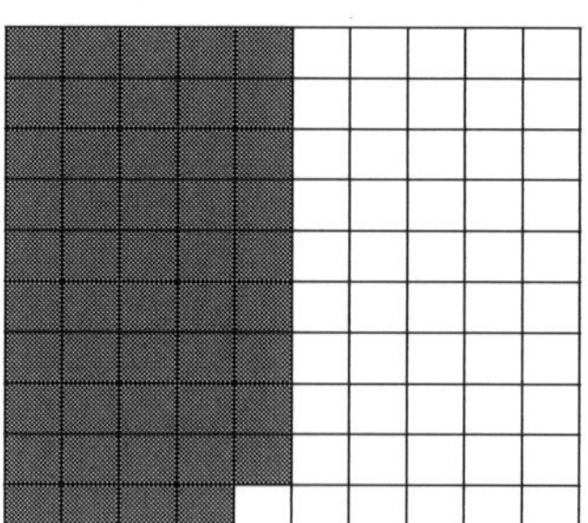

The 100 squares in each diagram represent 1 whole. What is the difference between the values represented by the shaded parts of each model?

____

4.NF.C.6

15. Which two decimals are greater than $\frac{1}{2}$, but less than $\frac{3}{4}$?

A. 0.5 and 0.72 **B.** 0.67 and 0.71
C. 0.59 and 0.76 **D.** 0.57 and 0.75

(4.NF.C.7)

16. Mario, Lulu, and Leo have a total of $$1.00 in coins.

- Mario has 2 dimes, 3 nickels and 4 pennies
- Lulu has 1 dime, 6 nickels and 2 pennies
- Leo has 1 dime, 1 nickel and 4 pennies

Which inequality compares the amount of money Lulu has to the combined amount of money Mario and Leo have?

A. $0.42 > $0.58 **B.** $0.39 < $0.42
C. $0.58 > $0.42 **D.** $0.42 > $0.19

(4.NF.C.7)

17. Which two decimals are greater than $\frac{1}{4}$, but less than $\frac{3}{8}$?

A. 0.25 and 0.31 **B.** 0.27 and 0.45
C. 0.26 and 0.33 **D.** 0.24 and 0.36

(4.NF.C.7)

18. Which points on this number line represent decimals that are less than $\frac{3}{5}$ and greater than $\frac{1}{4}$?

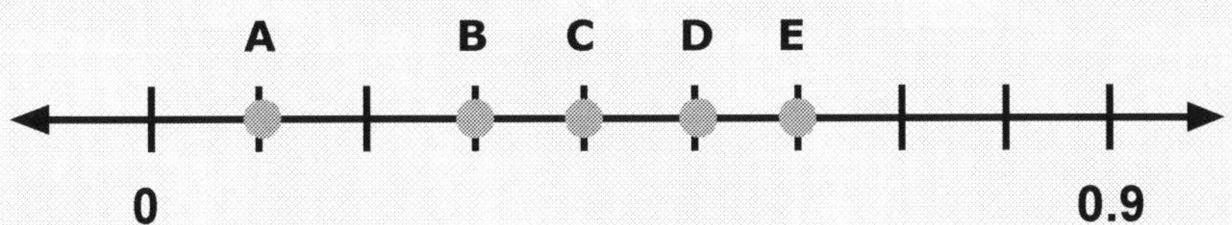

A. Points A, B, and C **B.** Points B, C, and D
C. Points B, C, D, and E **D.** Points D and E

(4.NF.C.7)

Ace Academic Publishing
ACHIEVING EXCELLENCE TOGETHER

19. Which point on this number line is greater than 0.8?

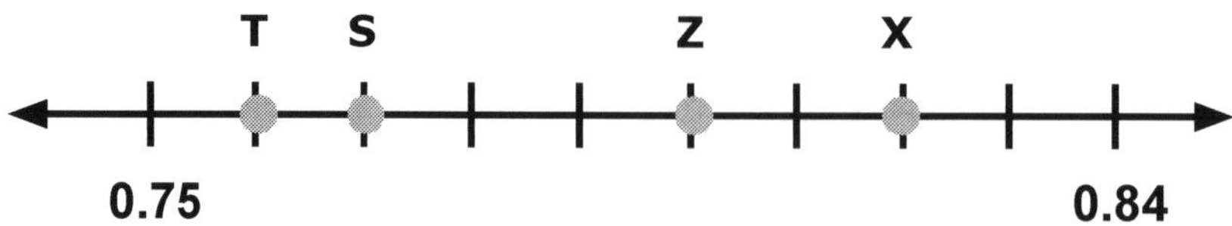

A. Point S **B.** Point T **C.** Point X **D.** Point Z

4.NF.C.7

20. What inequality can you write to compare Points A and B on this number line?

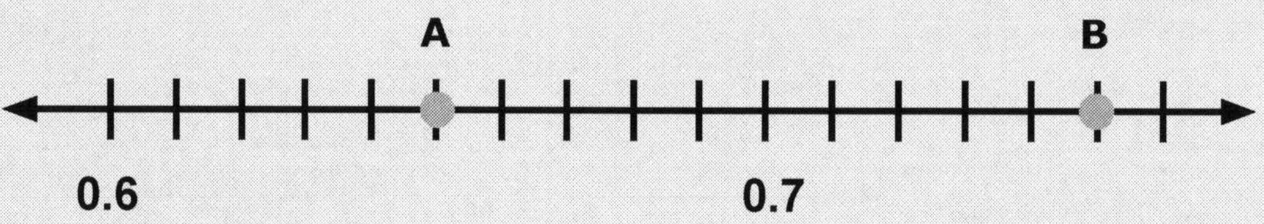

____

4.NF.C.7

1. Which fraction can be simplified into another equivalent fraction?

A. $\frac{4}{32}$ **B.** $\frac{8}{41}$ **C.** $\frac{7}{18}$ **D.** $\frac{9}{25}$

4.NF.C.7

2. Which shaded part of the fraction model shows a fraction equivalent to $\frac{1}{4}$.

A. 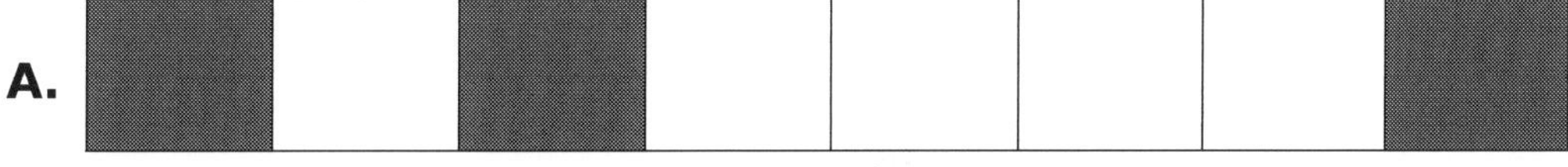

B. 

C. 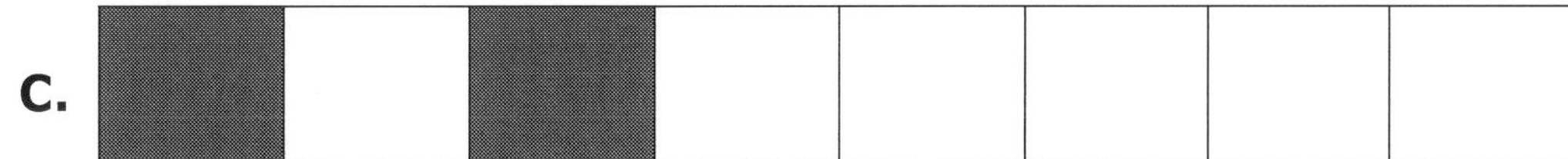

D. 

4.NF.A.1

3. Which fraction listed below is greater than $\frac{1}{2}$?

A. $\frac{2}{12}$ **B.** $\frac{3}{8}$ **C.** $\frac{2}{3}$ **D.** $\frac{1}{5}$

4.NF.A.2

CHAPTER REVIEW

4. Use the models below to compare:

$\frac{5}{8}$ and $\frac{1}{2}$

Compare using <, >, = and explain your reasoning.

____

____

____

(4.NF.A.2)

5. Anna is baking cakes. She uses $\frac{1}{6}$ cups of flour for the chocolate cake and $\frac{4}{6}$ cups of flour for the velvet cake. Which expression shows the number of cups of flour Anna used for the 2 cakes?

A. $\frac{1}{6}+\frac{1}{6}+\frac{1}{6}+\frac{1}{6}$

B. $\frac{1}{6}+\frac{1}{6}+\frac{1}{6}+\frac{1}{6}+\frac{1}{6}$

C. $\frac{1}{6}+\frac{1}{6}+\frac{1}{6}+\frac{1}{6}+\frac{1}{6}+\frac{1}{6}$

D. $\frac{1}{6}+\frac{1}{6}+\frac{1}{6}+\frac{1}{6}+\frac{1}{6}+\frac{1}{6}+\frac{1}{6}$

(4.NF.B.3)

6. Jerry has one bottle of soda. He pours Sam a glass with $\frac{3}{8}$ of the soda and another glass with $\frac{2}{8}$ of the soda for himself. Which expression represents the amount of soda remaining?

A. $\frac{1}{8}+\frac{1}{8}+\frac{1}{8}+\frac{1}{8}$

B. $\frac{1}{8}+\frac{1}{8}+\frac{1}{8}+\frac{1}{8}+\frac{1}{8}$

C. $\frac{1}{8}+\frac{1}{8}+\frac{1}{8}$

D. $\frac{1}{8}+\frac{1}{8}$

(4.NF.B.3)

7. Michael owns 12 pairs of pants and 24 shirts. One-third of his pants and $\frac{1}{4}$ of his shirts are blue.

Which equations can be used to represent the number of blue pants and blue shirts Michael owns?

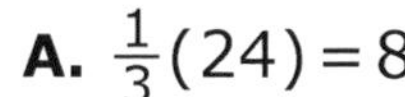

A. $\frac{1}{3}(24) = 8$
$\frac{1}{4}(12) = 3$

B. $\frac{1}{3}(36) = 12$
$\frac{1}{4}(36) = 9$

C. $\frac{1}{3} + 12 = 12\frac{1}{3}$
$\frac{1}{4} + 24 = 24\frac{1}{4}$

D. $\frac{1}{3}(12) = 4$
$\frac{1}{4}(24) = 6$

(4.NF.B.4)

8. Fill in the blanks to write an expression matching this model.

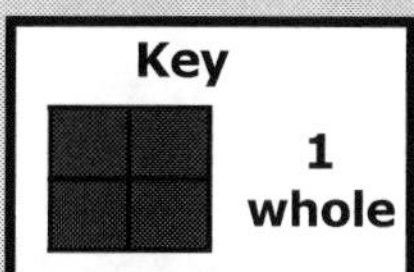

What fraction is represented by this model?

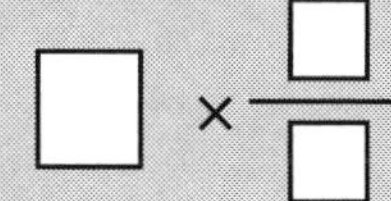

(4.NF.B.4)

9. Grace has a bowl with a capacity of $40\frac{5}{100}$ ounces. The capacity of Evan's bowl is 10 times this amount. Which fraction represents the capacity of Evan's bowl?

A. $50\frac{50}{100}$ ounces

B. $400\frac{5}{100}$ ounces

C. $40\frac{50}{100}$ ounces

D. $400\frac{50}{100}$ ounces

(4.NF.C.5)

10. The points on this number line represent two fractions.

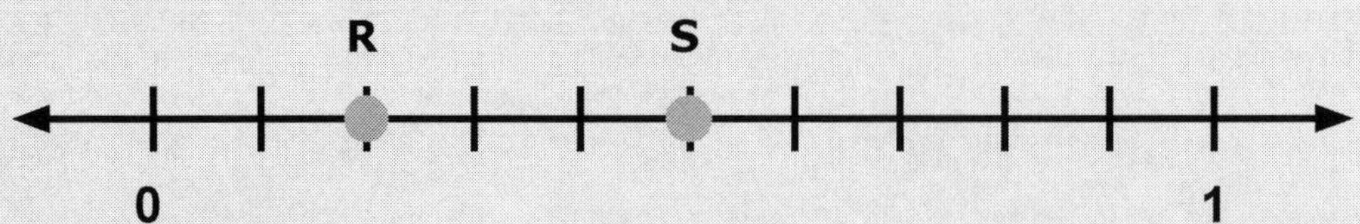

Which fraction is equivalent to the sum of the values at Point **R** and Point **S**?

A. $\frac{7}{100}$ **B.** $\frac{8}{100}$ **C.** $\frac{80}{100}$ **D.** $\frac{70}{100}$

4.NF.C.5

11. Jerome has written $\frac{6}{10}$ of his book report.

What is an equivalent fraction be if the denominator is 100?

____

4.NF.C.5

12. Point X on this number line represents a fraction.

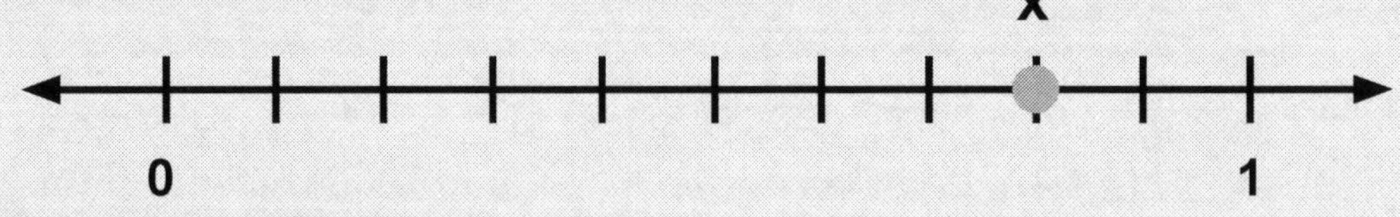

If $\frac{10}{100}$ is added to the fraction at Point X, what is the new fraction?

____

4.NF.C.5

13. Ping's pencil is $\frac{8}{10}$ inches longer than Riley's pencil. Riley's pencil is $9\frac{3}{10}$ inches long. How long is Ping's pencil?

A. 9.3 inches
B. 8.5 inches
C. 9.1 inches
D. 10.1 inches

(4.NF.C.6)

14. Sabra and Pari are writing a book. Paola has written $\frac{35}{100}$ of the book. Rohan has written $\frac{5}{10}$ of the book. Write an equation that gives the decimal that represents the fraction of the book they have written.

____

(4.NF.C.6)

15. This table shows the fraction out of 100 students who have the favorite colors listed in the table.

Color	Fraction
Red	$\frac{15}{100}$
Blue	$\frac{28}{100}$
Yellow	$\frac{8}{100}$
Green	?
Pink	$\frac{19}{100}$

The number of students who prefer green is twice the number of students who prefer red.

Which decimal represents the number of students who prefer red?

____

(4.NF.C.6)

CHAPTER REVIEW

CHAPTER REVIEW

16. Han is using this diagram to model a fraction.

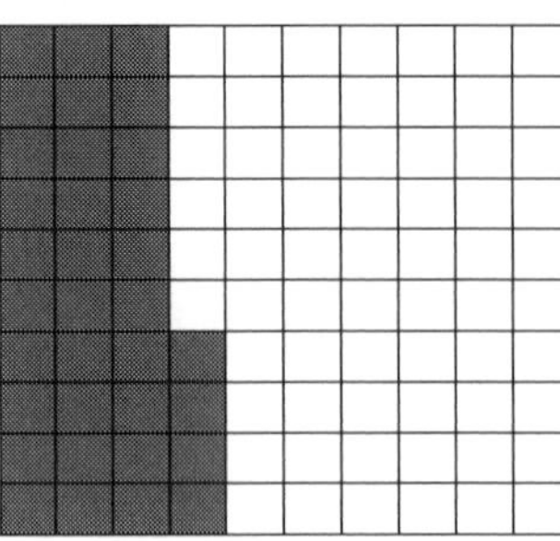

The 100 squares in this diagram represent 1 whole. Which decimal is represented by the shaded part of the model?

____

4.NF.C.6

17. Which sign makes this inequality true?

1.85   1.9

4.NF.C.7

18. Umar has 3 dimes, 6 nickels, and 8 pennies. Zane has \$0.79. Which inequality correctly compares the amount of money Umar has to the amount of money Zane has?

A. \$0.79 > \$0.68   **B.** \$0.79 < \$0.68

C. \$0.83 > \$0.79   **D.** \$0.83 < \$0.79

4.NF.C.7

19. Zack has 3 quarters and 3 pennies. Lara has \$0.14. Which inequality correctly compares the amount of money Umar has to the amount of money Zane has?

A. \$0.78 < \$0.14   **B.** \$0.14 < \$0.64

C. \$0.14 > \$0.64   **D.** \$0.78 > \$0.14

4.NF.C.7

20. Terry and Ivan are measuring the length of two pencils. Terry's pencil is 18.4 centimeters long. Ivan's pencil is 6.5 centimeters longer than Terry's pencil. What inequality could be used to compare the lengths of these pencils?

____

(4.NF.C.7)

NUMBER & OPERATIONS – FRACTIONS

EXTRA PRACTICE

1. Circle all the fractions that are equivalent to the fraction at the **X** on the number line.

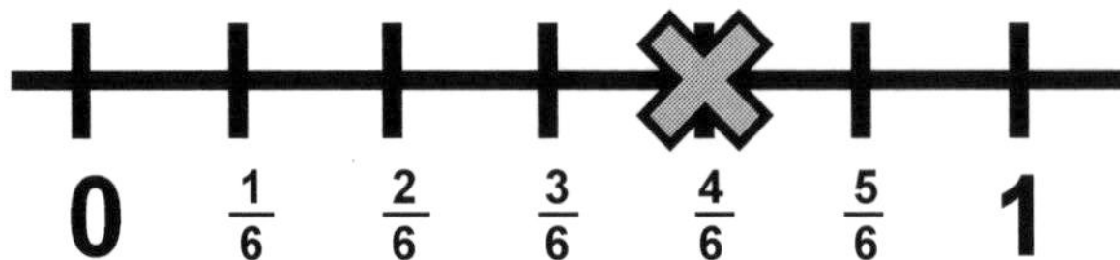

$\frac{5}{6}$ $\frac{8}{10}$ $\frac{8}{12}$ $\frac{2}{3}$ $\frac{3}{4}$ $\frac{12}{18}$

(4.NF.A.1)

2. Susan is counting her collection of shells. Out of 30 shells, 5 are black. She adds 6 more shells to her collection with the same fraction of black shells. How many black shells did she add to her collection? Exvplain how you found your answer.

____

____

____

(4.NF.A.1)

3. Which statement is true about $\frac{1}{4}$ and $\frac{3}{7}$?

A. $\frac{1}{4}$ is smaller because $\frac{1}{4}$ is closer to 0 and $\frac{3}{7}$ is closer to $\frac{1}{2}$.

B. $\frac{1}{7}$ is larger than $\frac{3}{4}$ because $7 > 4$.

C. $\frac{1}{4}$ is equal to $\frac{3}{7}$ because both fractions are close to 0.

D. $\frac{1}{4}$ is larger than $\frac{3}{7}$ because fourth-sized pieces are larger than seventh-sized pieces.

(4.NF.A.2)

4. Marco's teacher told him to spend $\frac{3}{5}$ of an hour on homework. He spent $\frac{4}{6}$ of an hour on his homework. Did Marco spend the required time on his homework? Explain your reasoning.

____

____

____

(4.NF.A.2)

EXTRA PRACTICE

5. Lisa is studying for a math test. She spends $\frac{3}{4}$ of an hour studying on Monday, $\frac{1}{4}$ of an hour studying on Tuesday, $\frac{3}{4}$ of an hour studying on Wednesday, and $\frac{2}{4}$ of an hour studying on Thursday. Does this number line to show the total amount of time Lisa spends studying for her test? Explain why.

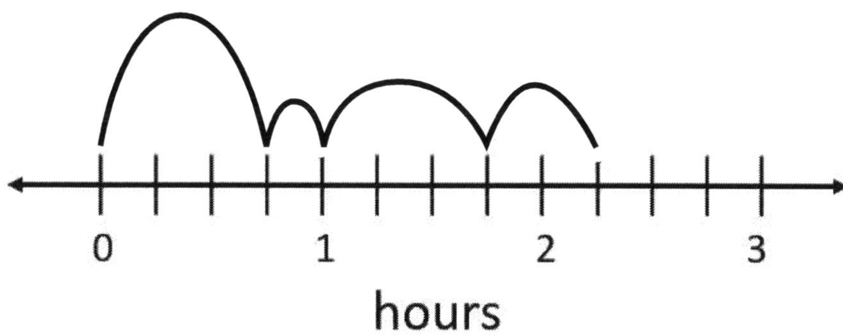

____

____

____

(4.NF.B.3)

EXTRA PRACTICE

6. Mrs. Bartley bought a pizza for dinner. She shared the pizza with her children. She ate $\frac{2}{10}$ of the pizza and there was $\frac{1}{10}$ of the pizza remaining after the children were finished eating. Which expression represents the fraction of pizza Mrs. Bartley's children ate?

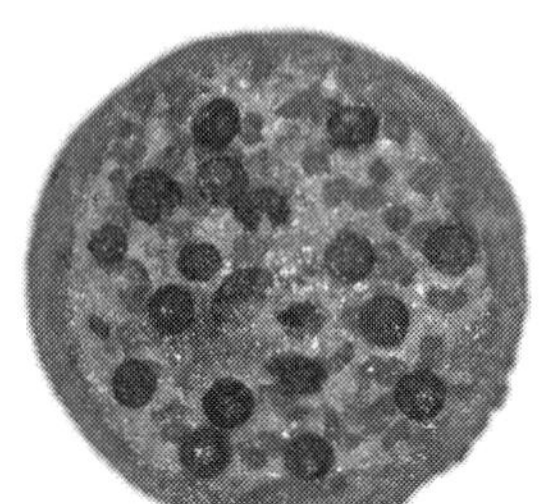

A. $\frac{1}{10}+\frac{1}{10}+\frac{1}{10}+\frac{1}{10}+\frac{1}{10}$

B. $\frac{1}{10}+\frac{1}{10}$

C. $\frac{1}{10}+\frac{1}{10}+\frac{1}{10}+\frac{1}{10}+\frac{1}{10}+\frac{1}{10}+\frac{1}{10}$

D. $\frac{1}{10}+\frac{1}{10}+\frac{1}{10}+\frac{1}{10}+\frac{1}{10}+\frac{1}{10}+\frac{1}{10}+\frac{1}{10}+\frac{1}{10}+\frac{1}{10}$

(4.NF.B.3)

7. John, Tyler, and Erik walk the same distance every day. This diagram shows the number of miles they walk each day.

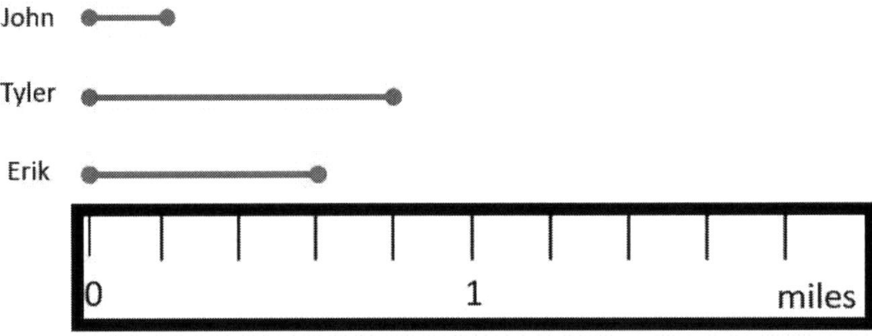

How would you determine the total number of miles John, Tyler, and Erik walk in 7 days?

____

____

____

(4.NF.B.4)

8. Waymon and Tessa each create a model to represent the expression $5 \times \frac{1}{2}$.

Waymon's Model

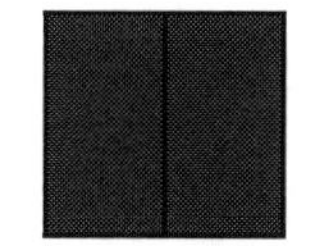
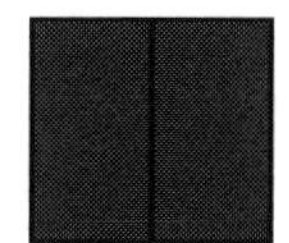

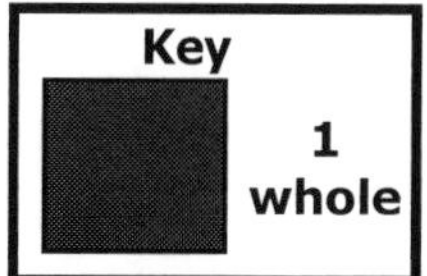

Tessa's Model

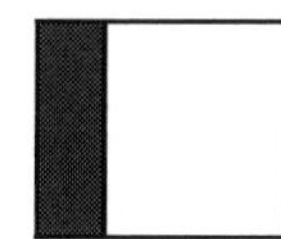
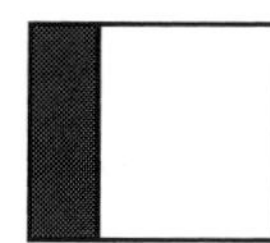
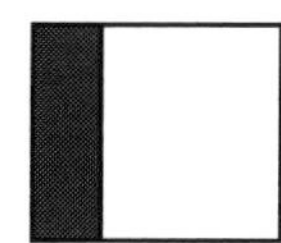
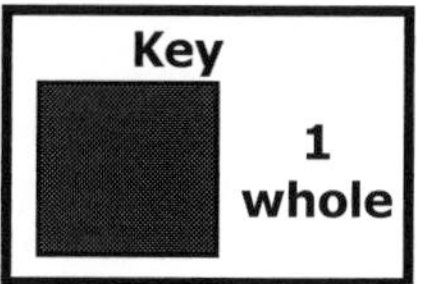

Are the models Waymon and Tessa create correct? Explain your reasoning.

____

____

____

(4.NF.B.4)

9. A bag contains red, blue, green, and black marbles. This table shows the portion of each marble, represented as a fraction.

Marble Color	Fraction of Bag
Red	$\frac{3}{10}$
Blue	$\frac{10}{100}$
Yellow	$\frac{20}{100}$
Green	$\frac{4}{10}$

Write an expression to represent the total number of blue and black marbles.

____

(4.NF.B.5)

EXTRA PRACTICE

EXTRA PRACTICE

10. Kindra uses pennies to estimate the length of this pencil.

Each penny is $\frac{8}{10}$ inches wide.
Does this expression to represent the length of the pencils?

$$\frac{80}{100} + \frac{80}{100} + \frac{80}{100} + \frac{80}{100} + \frac{80}{100} + \frac{80}{100}$$

Explain your reasoning.

____

____

____

(4.NF.B.5)

11. Han is using these models to represent equivalent fractions.

Model A **Model B**

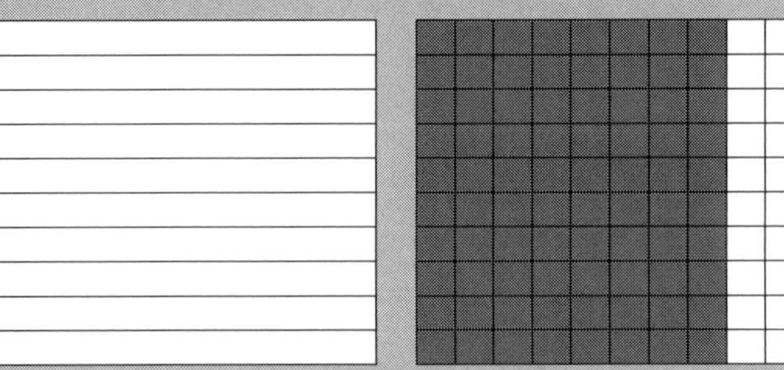

How many parts should be shaded in Model A to represent a fraction equivalent to the one shown in Model B?

A. 80 **B.** 8 **C.** 2 **D.** 20

(4.NF.B.5)

EXTRA PRACTICE

12. Edward's snake is 8 centimeters shorter than Jessica's snake. Jessica's snake is $54\frac{7}{10}$ centimeters long. Edward determines the length of his snake is $46\frac{70}{100}$ centimeters. Why is he correct?

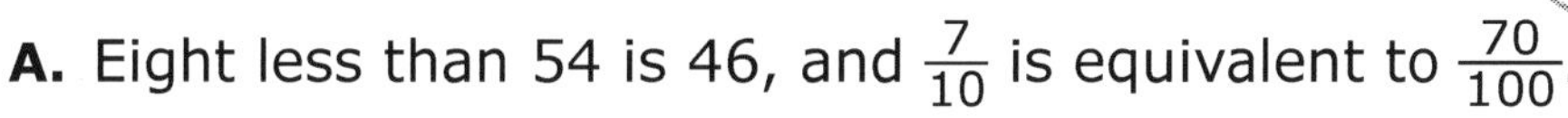

A. Eight less than 54 is 46, and $\frac{7}{10}$ is equivalent to $\frac{70}{100}$.

B. Edward's snake is not as long as Jessica's snake.

C. The sum of 46 and 8 is 54, and 7 is less than 70.

D. The fraction $\frac{70}{100}$ is not in lowest terms.

(4.NF.B.5)

13. Write out the steps you would use to convert $\frac{87}{10}$ to a decimal.

____

____

____

(4.NF.C.6)

14. Taj determines the distance between Point A and Point C on this number line is 0.13.

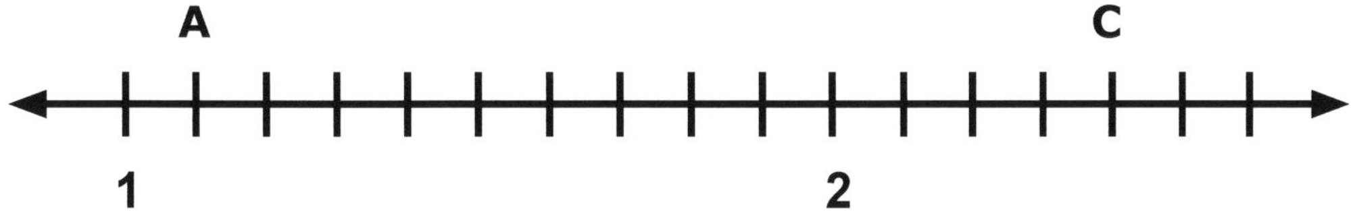

Do you agree with Taj? Explain your reasoning.

____

____

____

(4.NF.C.6)

EXTRA PRACTICE

15. Mr. Johnson is cooking a soup. He fills $\frac{3}{10}$ of the pot with carrots, $\frac{4}{10}$ of the pot with broth, and $\frac{1}{10}$ of the pot with chicken. Which decimal represents the amount of space left in the pot?

A. 0.2 **B.** 0.4 **C.** 0.6 **D.** 0.8

(4.NF.C.6)

16. Why is $\frac{54}{100}$ equivalent to 0.54?

A. Fifty-four hundredths is equivalent to $\frac{54}{1000}$.
B. Fifty-four divided by 100 is 0.54.
C. Fifty-four hundredths is greater than 0.50.
D. The fraction $\frac{54}{100}$ is not expressed in lowest terms.

(4.NF.C.6)

17. Which decimal represents the additional shaded squares needed for the value of the shaded area of this model to be equivalent to 0.92? Explain your reasoning.

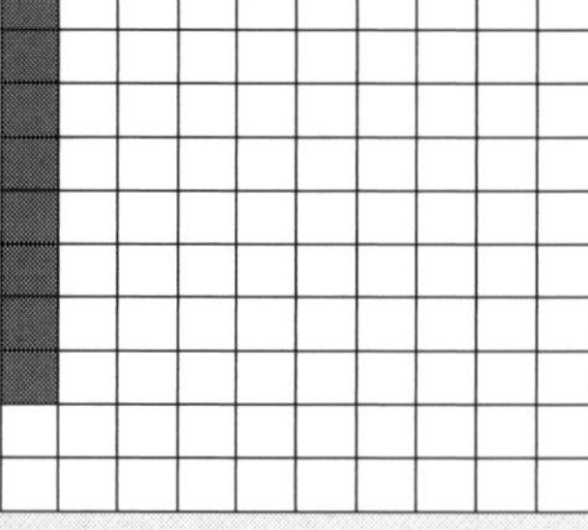

____

____

____

(4.NF.C.7)

18. Jack marks two decimals on a number line. One decimal is 0.96. The other decimal is 0.37 greater than 0.96.

What is the value of the larger decimal?

____

(4.NF.C.7)

19. Nate and George are measuring the height of two plants. George's plant is 1.84 feet tall. Nate's plant is 0.18 feet shorter than George's plant. What inequality can be used to compare the height of their plants?

____

(4.NF.C.7)

20. Winnie eats $\frac{1}{4}$ of this bucket of popcorn.

Michael eats $\frac{1}{2}$ of this bucket of popcorn.

Winnie knows $\frac{1}{4}$ is equivalent to 0.25, and $\frac{1}{2}$ is equivalent to 0.50 Winnie thinks Michael eats more popcorn. Do you agree? Explain your reasoning.

____

____

____

(4.NF.C.7)

Success

MEASUREMENT & DATA

4

PROBLEM SOLVING - CONVERSION OF MEASUREMENTS 107

GRAPHS AND DATA INTERPRETATION 111

GEOMETRY - CONCEPTS OF ANGLES 121

CHAPTER REVIEW 128

EXTRA PRACTICE 134

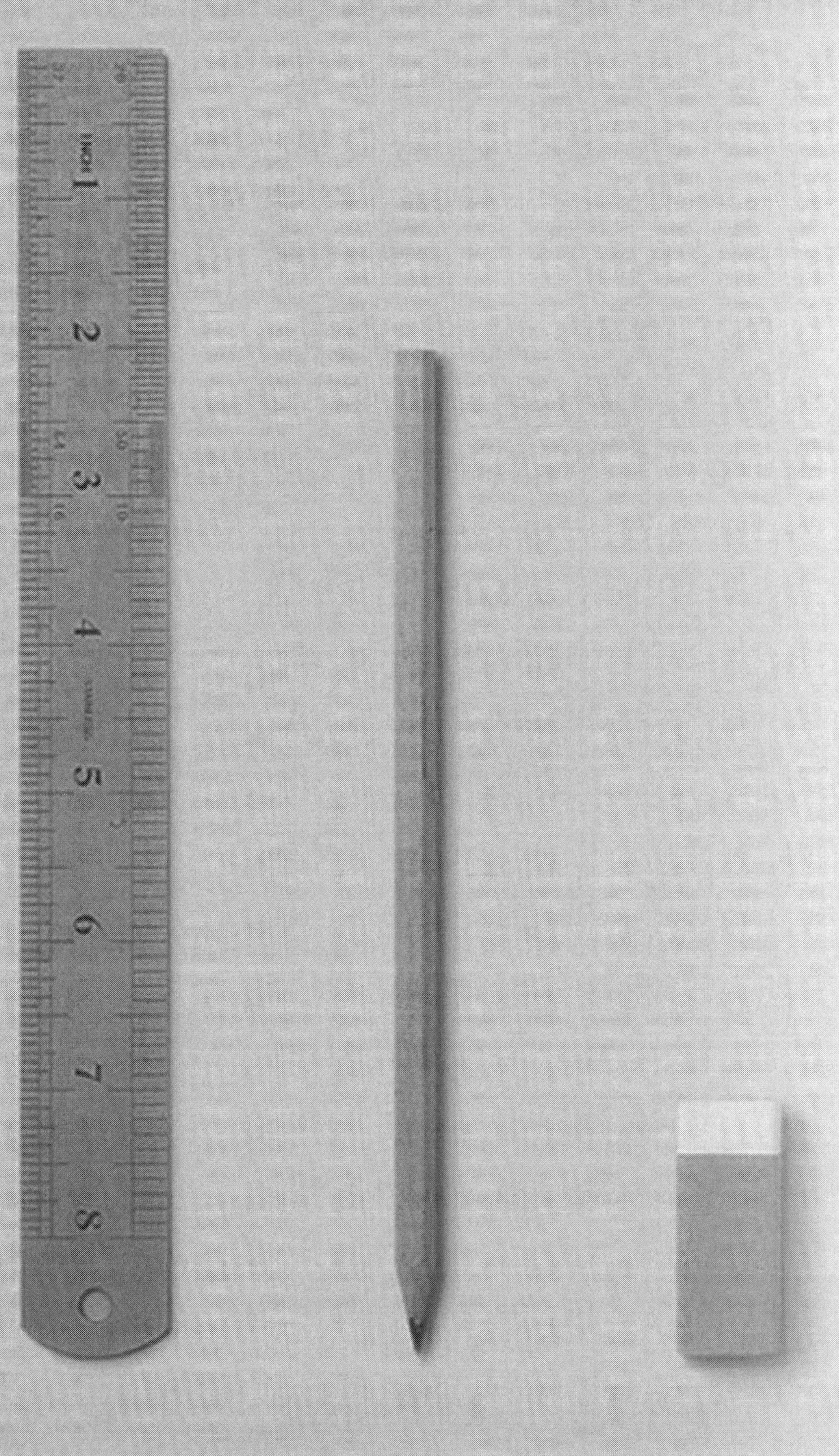
1
2
3
4
5
6
7
8

1. Which is the best estimate for the length of a golf club? 

A. 35 inches B. 35 yards C. 35 feet D. 35 miles

4.MD.A.1

2. Which is the best estimate for the height of a telephone pole?

A. 18 yards B. 18 inches C. 18 feet D. 18 miles

4.MD.A.1

3. How many yards is equal to 72 inches?

A. 2 B. 4 C. 3 D. 1

4.MD.A.1

4. Which quantity is greatest: 1 ton, 2,000 ounces or 230 pounds?

A. 1 ton B. 2,000 ounces C. 230 pounds D. they are all equal

4.MD.A.1

5. How many pints is one gallon?

A. 4 B. 6 C. 8 D. 10

4.MD.A.1

6. Which is the greatest value?

1 pint, 0.5 quarts or 3 cups?

A. 1 pint B. 3 cups C. 0.5 quarts D. they are equal

4.MD.A.1

7. Which quantity is greatest: 2 pints, 1 cup, 1 tablespoon or 10 teaspoons?

A. 1 tablespoon B. 10 teaspoons C. 1 cup D. 2 pints

4.MD.A.1

8. Del bought a cookie sheet for $8.07. He gave $8.55 to the cashier. How much change did Del receive?

A. $0.48 **B.** $0.55 **C.** $0.38 **D.** $0.42

4.MD.A.2

9. Marine purchased a bottle of nail polish costing $2.73. She gave the cashier $7.50. How much change did the cashier give to Marine?

A. $4.07 **B.** $4.77 **C.** $2.77 **D.** $5.37

4.MD.A.2

10. Aubrey purchased a bowl of chicken noodle soup for $3.77. She gave the cashier $6.25. How much change did Aubrey receive?

A. $0.48 **B.** $1.48 **C.** $2.48 **D.** $3.48

4.MD.A.2

11. Stephanie bought a roll of wrapping paper costing $3.64. She gave the cashier $5.00. How much change did the cashier give back to Stephanie?

A. $2.36 **B.** $1.36 **C.** $0.36 **D.** $3.64

4.MD.A.2

12. Heddy purchased a cherry pie for $5.24. She gave the cashier $5.50. How much change did Heddy receive?

A. $0.36 **B.** $2.26 **C.** $1.26 **D.** $0.26

4.MD.A.2

13. A field day at the elementary school began at 9 AM and finished at 2:30 PM. How long was the field day?

A. 5 hours 30 minutes **B.** 7 hours 30 minutes
C. 3 Hours 30 minutes **D.** 4 hours 30 minutes

4.MD.A.2

14. Which of the following volumes is the largest?

A. 5 gallons **B.** 16 quarts **C.** 25 cups **D.** 320 ounces

4.MD.A.2

15. What is the perimeter of the rectangle? ________ cm

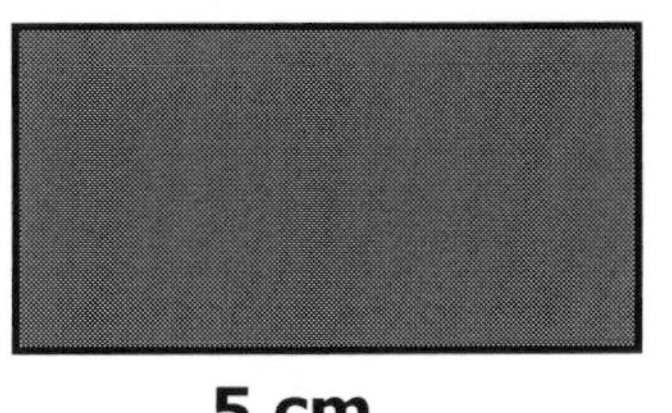

A. 25 **B.** 20
C. 18 **D.** 9

4.MD.A.3

16. What is the perimeter of the rectangle? ________ yds

A. 6 **B.** 8
C. 4 **D.** 5

4.MD.A.3

17. What is the perimeter of the rectangle? ________ cm

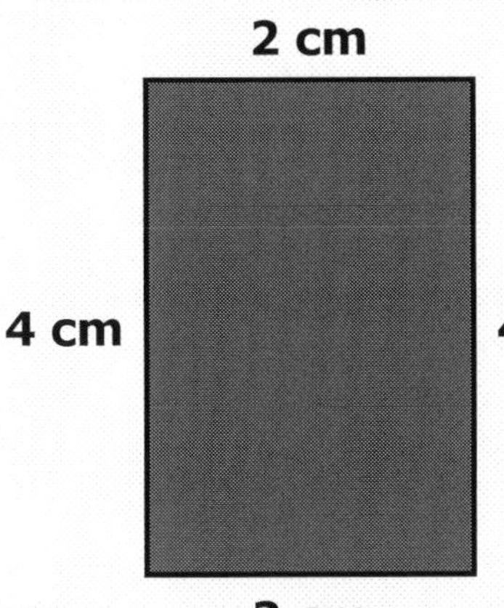

A. 6

B. 12

C. 8

D. 10

4.MD.A.3

18. What is the perimeter of the rectangle? ________ ft

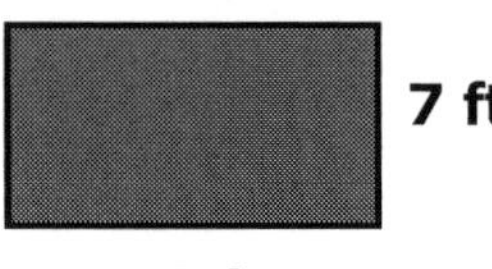

A. 14 **B.** 26

C. 38 **D.** 84

(4.MD.A.3)

19. What is the perimeter of the shape? ________ cm

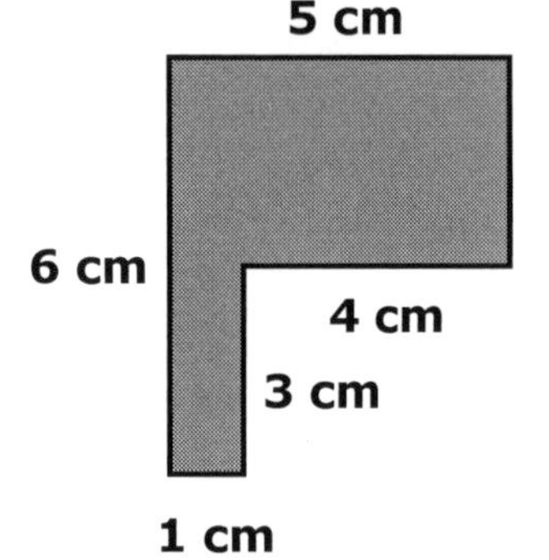

A. 11 **B.** 26

C. 12 **D.** 22

(4.MD.A.3)

20. What is the perimeter of the shape? ________ in

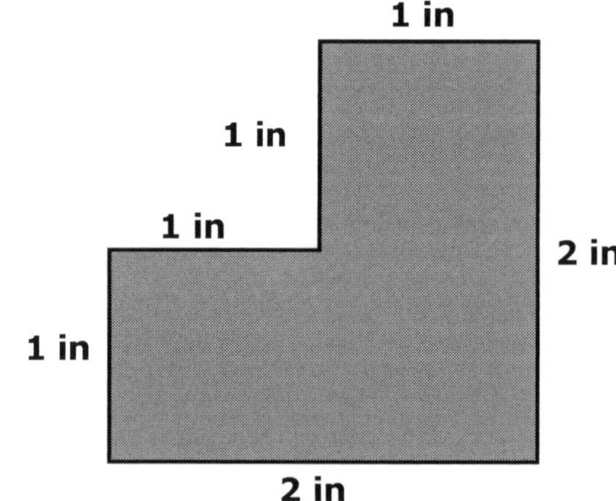

(4.MD.A.3)

1. Use the data in the table to finish the line plot below.

Length of Pencils (cm)	
Pencil Length	Number of Pencils
7 1/2	5
8	6
8 1/2	3
9 1/2	2

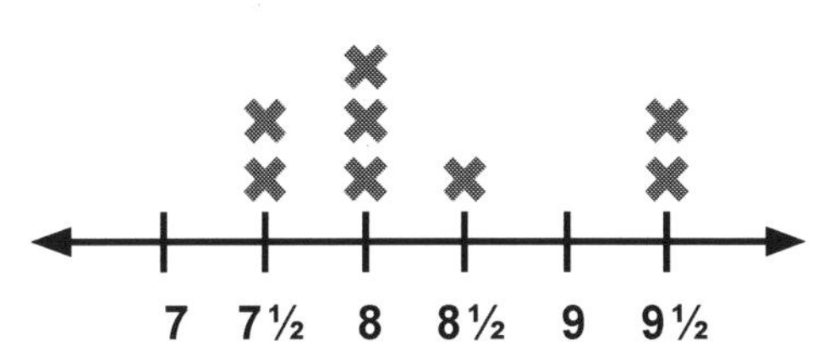

4.MD.B.4

2. Use the data in the table to finish the line plot below.

Cups of Fruit Used in Pie	
Number of Cups of Fruit	Number of Pies made
3/4	4
1	3
1 1/4	1
1 1/2	6

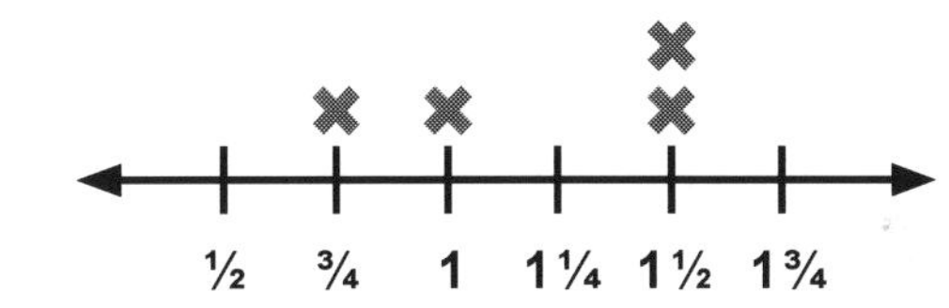

Number of Pies made with varying cups of Fruit

4.MD.B.4

3. Use the data in the table to finish the line plot below.

Baking Cookies	
Number of Batches Baked	Number of Bakers
3/4	5
1	3
1 1/4	9
1 1/2	6
1 3/4	8
2	2

Number of Cookie Batches Baked

4.MD.B.4

4. Use the data in the tally chart to create a line plot.

Hours Spent Reading	
Number of Hours	Number of Students
1	~~||||~~ |||
1 1/8	||||
1 1/4	||||
1 3/8	~~||||~~ |
1 1/2	~~||||~~
1 5/8	||

(4.MD.B.4)

5. Use the data in the table to create a line plot.

Hours Spent at Soccer Practice	
Number of Hours	Number of Players
3/4	4
1	2
1 1/4	3
1 1/2	3
1 3/4	1
2	1

(4.MD.B.4)

6. Use the data in the table to create a line plot.

Student Height	
Height (inches)	Number of Students
51	3
51 ¼	4
51 ½	8
51 ¾	2
52	6
52 ¼	1

(4.MD.B.4)

7. Students in Mr. Johnson's class selected apple juice cups based on how many ounces were in each cup. He counted the number of cups of apple juice his second-grade students selected. He presented the data on this line plot.

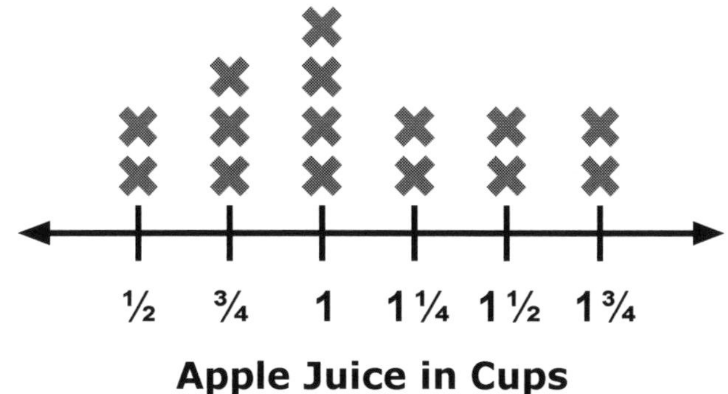

How many students had less than $1\frac{1}{4}$ cups of apple juice?

____

(4.MD.B.4)

GRAPHS AND DATA INTERPRETATION

8. Use the data in this table to complete the line plot.

Hours Studying				
Number of Hours Studied	3/4	1	1 1/4	1 1/2
Number of Students	5	7	8	6

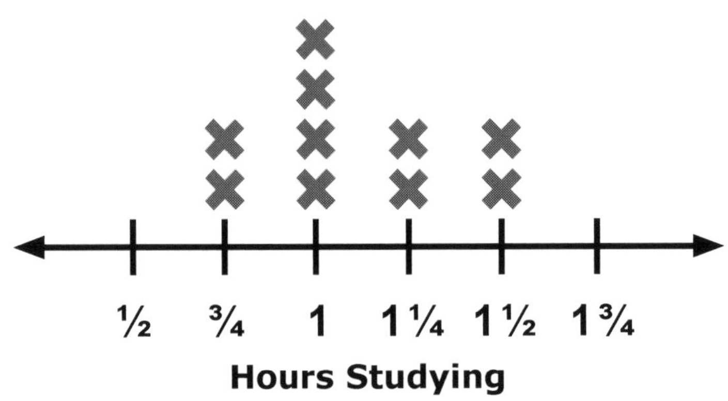

4.MD.B.4

9. Use this list of data to complete the line plot.

Hours of Exercise: $\frac{1}{2}$, 1, $1\frac{1}{4}$, $1\frac{1}{2}$, $1\frac{3}{4}$, $\frac{1}{2}$, $\frac{1}{2}$, 1, 1, 1, $1\frac{1}{4}$, $1\frac{1}{4}$, $1\frac{3}{4}$

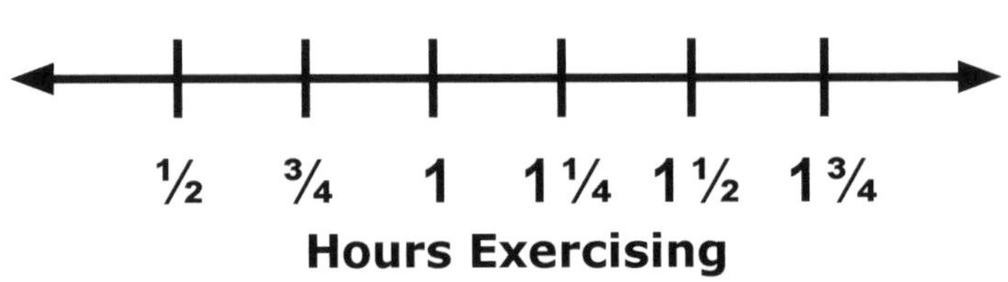

4.MD.B.4

GRAPHS AND DATA INTERPRETATION

10. Use the data in the table to complete the line plot.

Length of Worms					
Worm Length (inches)	3 3/4	4	4 1/4	4 1/2	5
Number of Worms	3	4	2	5	2

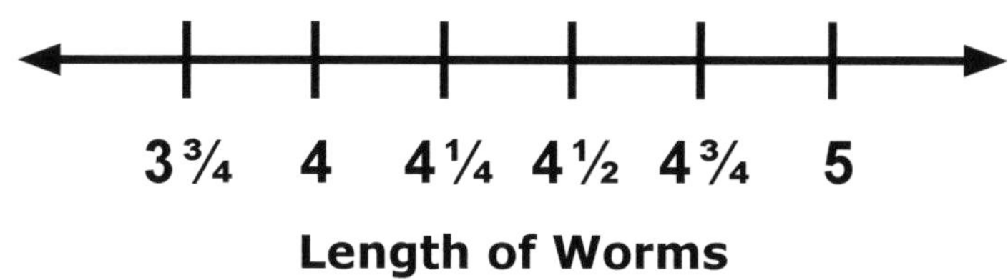

Length of Worms

4.MD.B.4

11. Use this list of data to complete the line plot.

Cups of Sugar per Day: $\frac{1}{4}$, $\frac{1}{4}$, $\frac{3}{4}$, $\frac{1}{2}$, $\frac{1}{2}$, $\frac{1}{4}$, $\frac{1}{4}$, $\frac{1}{4}$, $\frac{1}{2}$, $\frac{1}{2}$

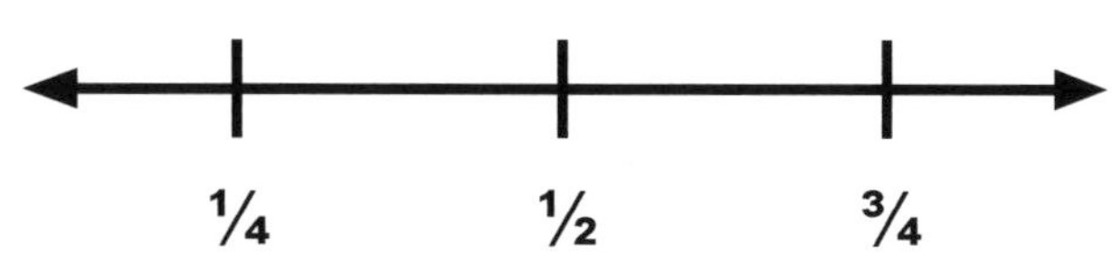

Cups of Sugar per Day

4.MD.B.4

GRAPHS AND DATA INTERPRETATION

12. Use the data in the table to create a line plot.

Foot Length					
Foot Length (inches)	$5\frac{1}{8}$	$5\frac{3}{8}$	$5\frac{1}{2}$	$5\frac{3}{4}$	6
Number of People	3	4	2	5	2

Foot Length

4.MD.B.4

13. Use the data in the table to create a line plot.

Dog Weight				
Dog Weight (lbs.)	$12\frac{7}{8}$	13	$13\frac{1}{4}$	$13\frac{1}{2}$
Number of Dogs	4	2	5	3

Dog Weigth (lb.)

4.MD.B.4

GRAPHS AND DATA INTERPRETATION

14. Use the data in this table to complete the line plot.

Amount of Buttons Used					
Buttons Used (cups)	3 3/4	4	4 1/4	4 1/2	5
Number of People	7	6	4	6	3

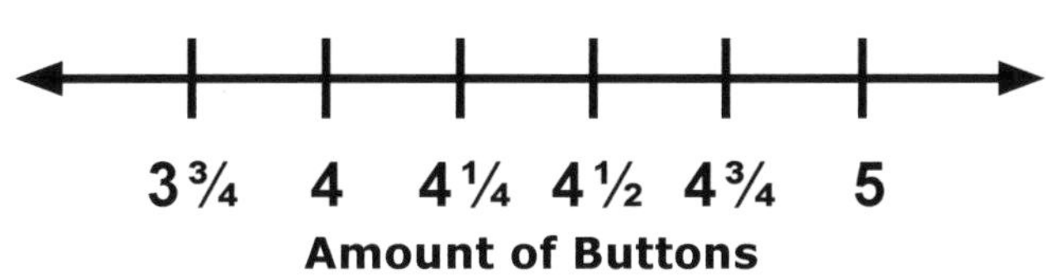

4.MD.B.4

15. Use the data in this table to complete the line plot.

Butterfly Length					
Length (inches)	3 3/4	4	4 1/4	4 1/2	5
Number of Butterflies	4	5	6	5	3

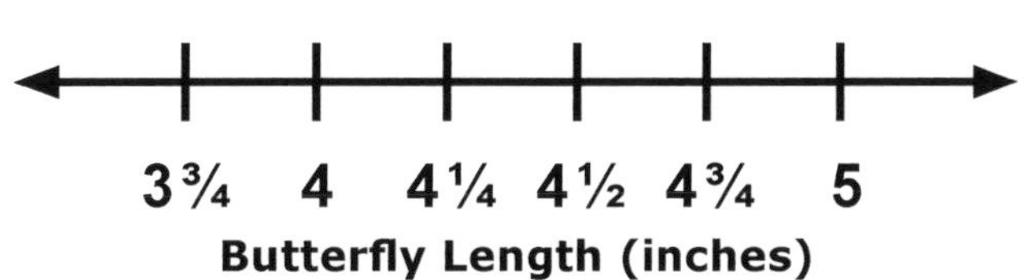

4.MD.B.4

GRAPHS AND DATA INTERPRETATION

16. Use this list of data to create a line plot.

Height of Tomato Plants (inches): $4\frac{1}{8}$, $4\frac{1}{2}$, $4\frac{3}{4}$, $4\frac{1}{2}$, $4\frac{1}{4}$, $4\frac{1}{8}$, $4\frac{3}{4}$, $4\frac{3}{4}$

Height of Tomato Plants (in.)

4.MD.B.4

17. Use the data in this table to create a line plot.

Ounces of Cologne								
2	1 ½	1 ¼	1 ½	1 ½	1	2	1 ¼	2

Ounces of Cologne

4.MD.B.4

18. In her cooking class, Ms. Stowe gave each student some hot chocolate mix. Use this data to create a line plot.

Hot Chocolate Mix	
Students	Tablespoons
Peter	2/4
Brandon	1/4
Lola	2/4
Symone	3/4
Enrique	2/4

Hot Chocolate Mix (tablespoons)

4.MD.B.4

19. Mrs. Agbomi volunteered to cook all the meals for the faculty on their camping retreat. She calculated how much propane fuel she would use during each meal and entered the amounts in the table below. Create a line plot to represent the amount of propane fuel needed.

Amount of Fuel	
Meal	Number of Cans
Breakfast 1	3/4
Lunch 1	3/4
Dinner 1	1
Breakfast 2	3/4
Lunch 2	2/4
Dinner 2	3/4
Breakfast 3	3/4

Amount of Fuel (cans)

4.MD.B.4

GRAPHS AND DATA INTERPRETATION

20. This table shows the amount of white paint on 6 palettes in an art class. Create a line plot from this data.

Amount of Paint	
Palette	Tablespoons
Palette 1	1
Palette 2	3/4
Palette 3	3/4
Palette 4	3/4
Palette 5	2/4
Palette 6	3/4

Amount of White Paint (tablespoons)

4.MD.B.4

1. What is the measurement of this angle?

A. 0°
B. 270°
C. 90°
D. 180°

(4.MD.C.5.A)

2. What is the measurement of this angle?

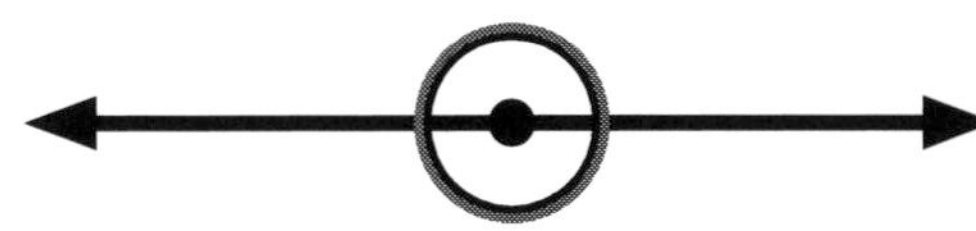

A. 180 degrees
B. 360 degrees
C. 270 degrees
D. 90 degrees

(4.MD.C.5.A)

3. What is the measure of this angle? Choose the best estimate.

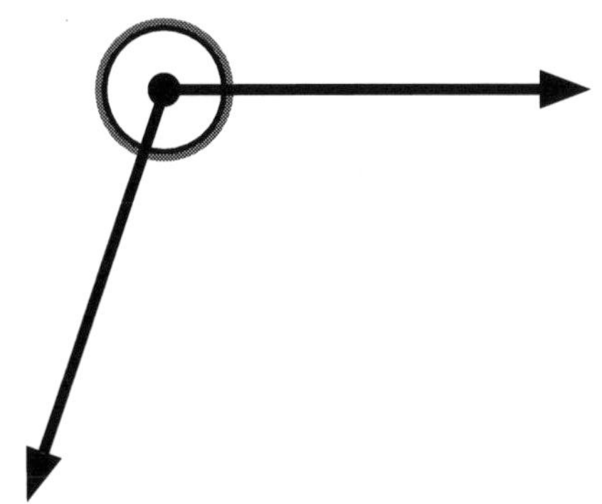

A. 70°
B. 360°
C. 90°
D. 120°

(4.MD.C.5.A)

GEOMETRY – CONCEPTS OF ANGLES

4. What fraction of a turn is this angle? ____________

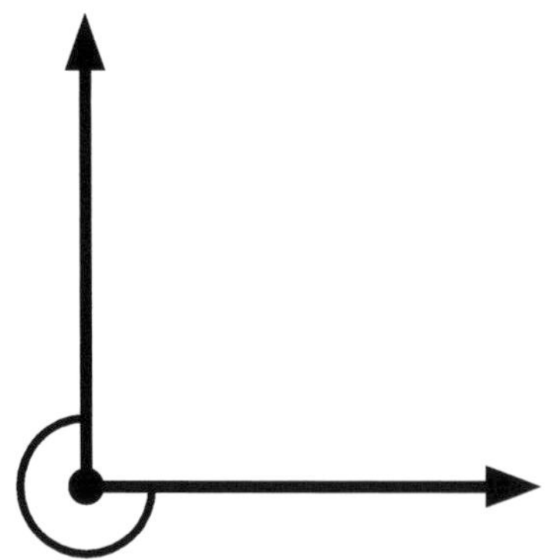

4.MD.C.5.A

5. Porsha turns the dial on her machine 8 times. Each time she turns the dial, she moves it 20 degrees. What is the total number of degrees she turns the dial?

____

4.MD.C.5.B

6. The minute hand on a clock moves 1 degree every 60 seconds. How many degrees has the minute hand moved after 600 seconds?

____

4.MD.C.5.B

7. The blades of a fan move 120 degrees each second. How many seconds does it take for the blades to turn a complete circle?

____

4.MD.C.5.B

8. What is the measure of this angle

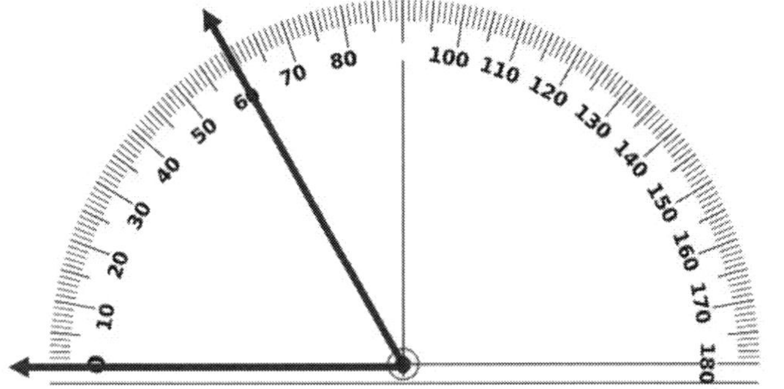

A. 60°
B. 0°
C. 180°
D. 90°

(4.MD.C.6)

9. What is the measure of this angle?

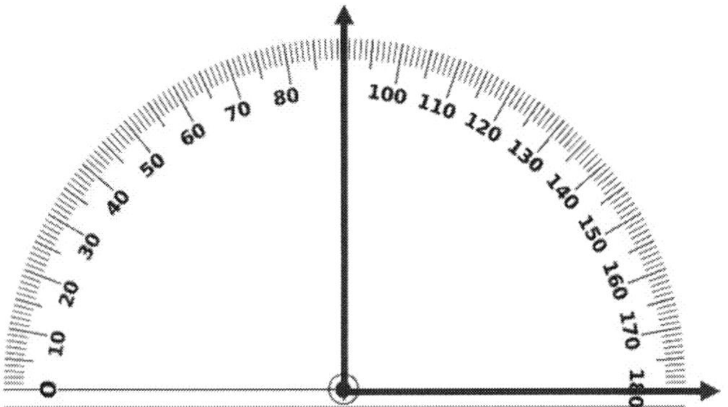

A. 0°
B. 30°
C. 90°
D. 80°

(4.MD.C.6)

10. What is the measure, in degree, of this angle?

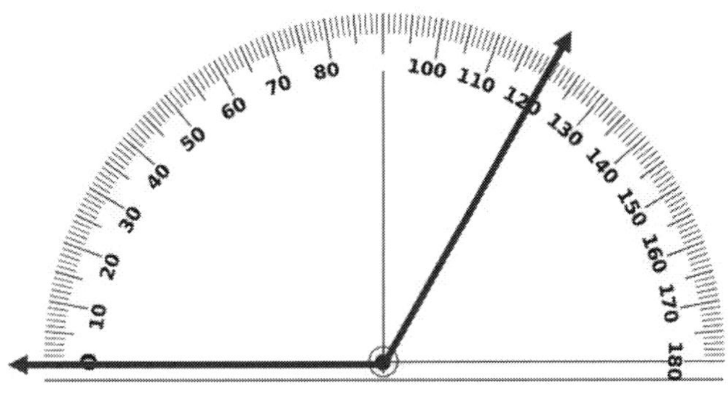

____

(4.MD.C.6)

GEOMETRY – CONCEPTS OF ANGLES

11. What is the measure of this angle?

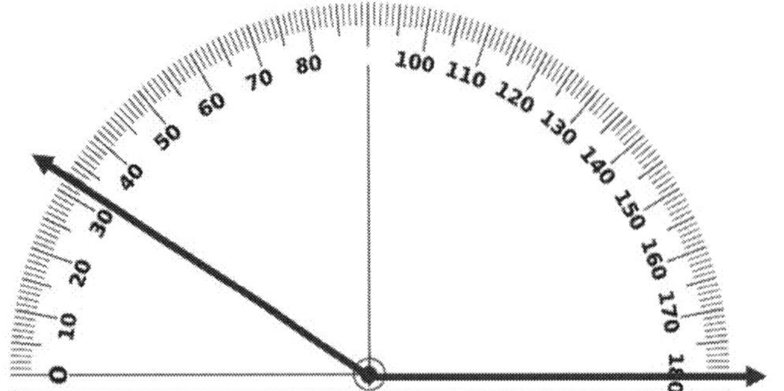

A. 150°
B. 33°
C. 180°
D. 147°

(4.MD.C.6)

12. What is the measure of this angle?

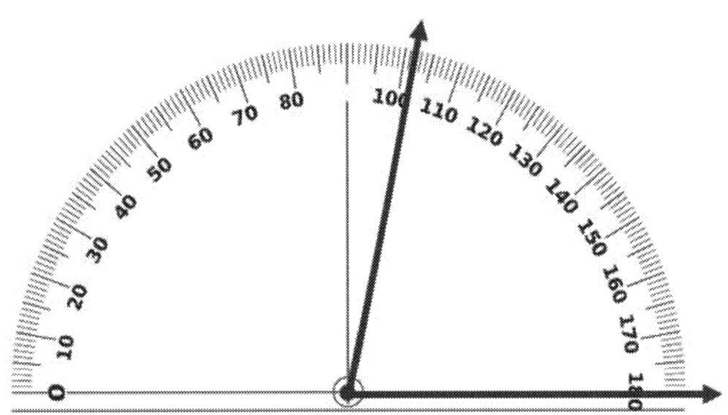

A. 88°
B. 102°
C. 78°
D. 180°

(4.MD.C.6)

13. What is the measure, in degrees, of this angle?

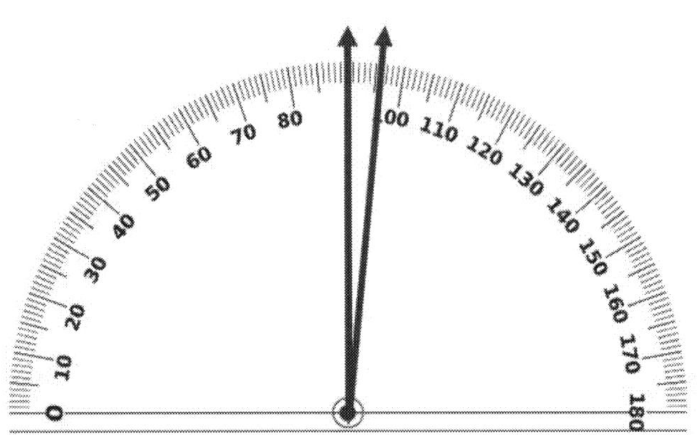

____

(4.MD.C.6)

14. What is the measure of this angle?

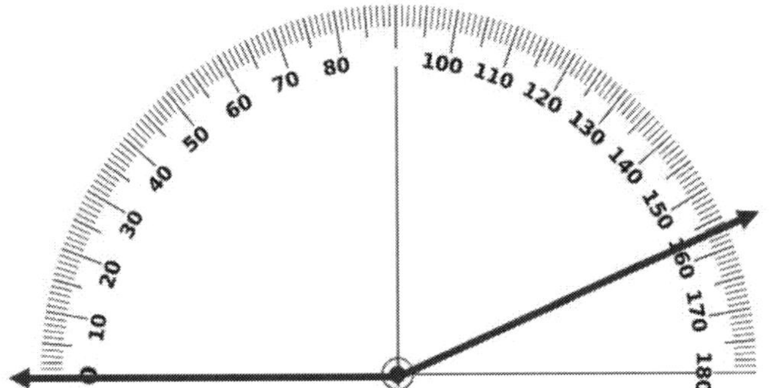

A. 120°

B. 150°

C. 157°

D. 180°

(4.MD.C.6)

15. What is the measure of angle *w*?

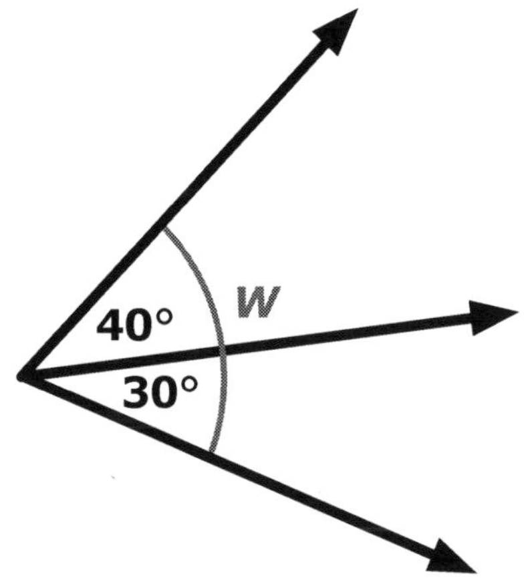

A. 90°

B. 10°

C. 70°

D. 80°

(4.MD.C.7)

16. What is the measure, in degrees, of angle *y*? ____________°

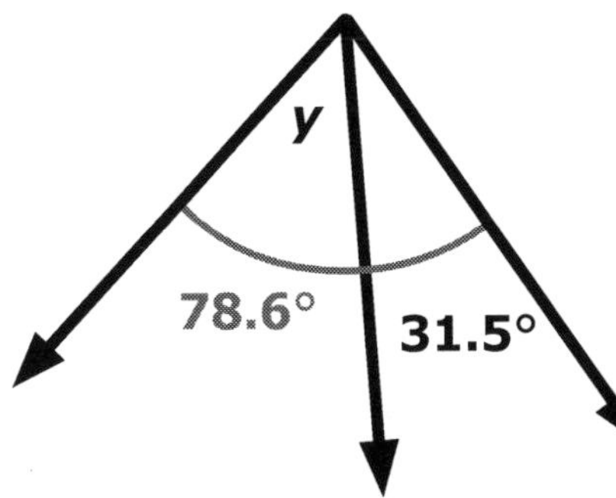

(4.MD.C.7)

GEOMETRY – CONCEPTS OF ANGLES

17. What is the measure angle *k*?

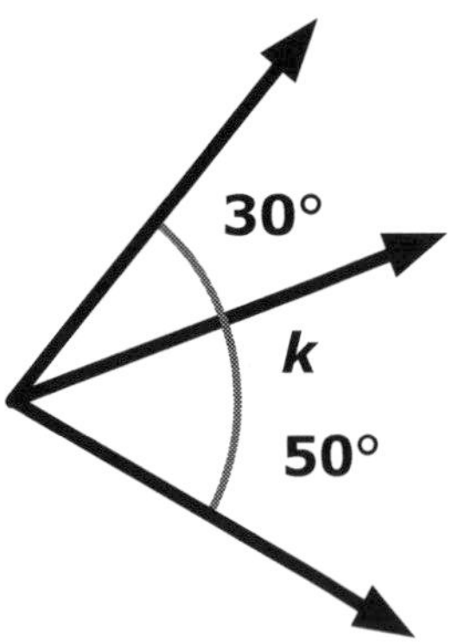

A. 90°
B. 80°
C. 60°
D. 50°

(4.MD.C.7)

18. What is the measure, in degrees, of angle *u*? ______________°

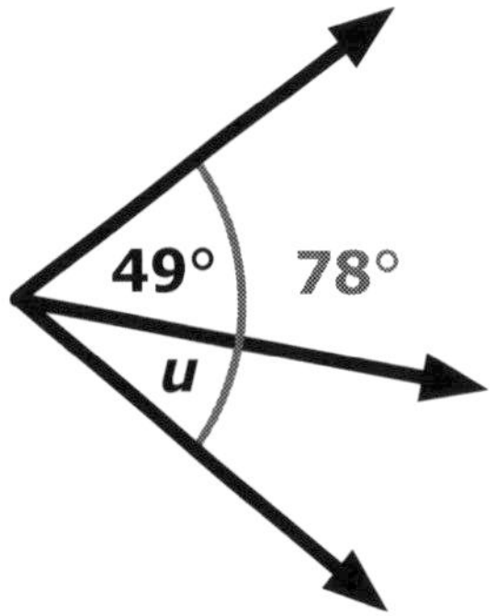

(4.MD.C.7)

19. Elias and Shawn are sharing a pizza cut into 5 large pieces. This model represents the pizza.

Elias and Shawn each take 2 pieces of pizza and decide to divide the last piece equally. Which expression can be used to determine the angle created by dividing the last piece of pizza into two equal parts?

A. $(360 \div 5)$ **B.** $(360 \div 2)$ **C.** $(360 \div 5) \times 2$ **D.** $(360 \div 5) \div 2$

(4.MD.C.7)

20. What is the measure, in degrees, of angle *w*? ____________°

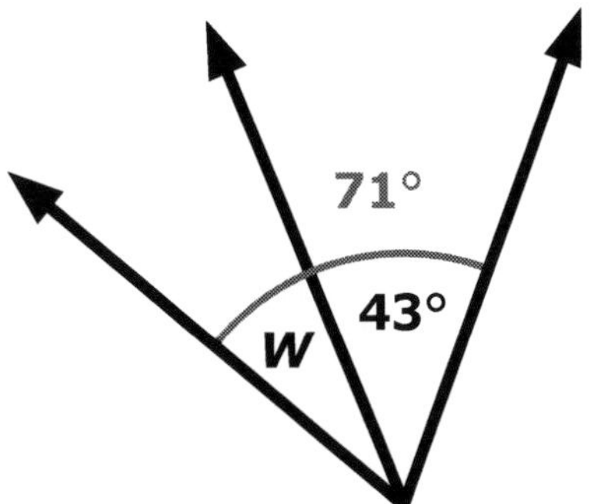

(4.MD.C.7)

21. Explain

MEASUREMENT & DATA

CHAPTER REVIEW

1. How many pounds are in 3 tons?

A. 2,000 B. 6,000 C. 4,000 D. 8,000

(4.MD.A.1)

2. Which is the best estimate for the length of a peanut?

A. 25 kilometers B. 25 meters
C. 25 centimeters D. 25 millimeters

(4.MD.A.1)

3. Which is the best estimate for the volume of a bathroom sink?

A. 8 milliliters B. 8 liters
C. 8 kiloliters D. 8 centiliters

(4.MD.A.1)

4. A 9-pack of plastic cups costs $4.23. What is the price of 1 pack?

A. $0.47 B. $0.49
C. $0.37 D. $0.31

(4.MD.A.2)

5. Thomas used 6 quarts of orange juice and 1 pint of soda to make punch for his party. During the party, Thomas' guests drank 3 quarts of the punch. How much punch is left?

A. 3 qt 1 pt B. 1 qt 1 pt
C. 5 qt 1 pt D. 4 qt 1 pt

(4.MD.A.2)

6. Mary used 2 tablespoons of sugar and 4 teaspoons of baking powder in a recipe. If there are 3 teaspoons in 1 tablespoon, how many tablespoons of these two ingredients did Mary use?

A. $2\frac{1}{2}$ tablespoons **B.** 4 tablespoons

C. $3\frac{1}{3}$ tablespoons **D.** $2\frac{2}{3}$ tablespoons

(4.MD.A.2)

7. What is the perimeter of the shape?

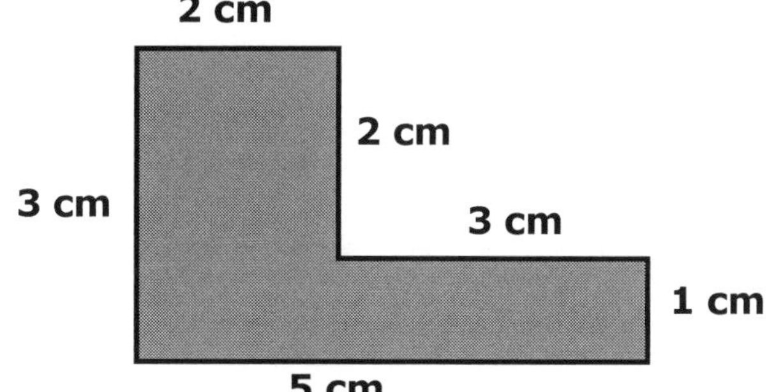

A. 10 cm
B. 8 cm
C. 16 cm
D. 15 cm

(4.MD.A.3)

8. What is the perimeter of the shape?

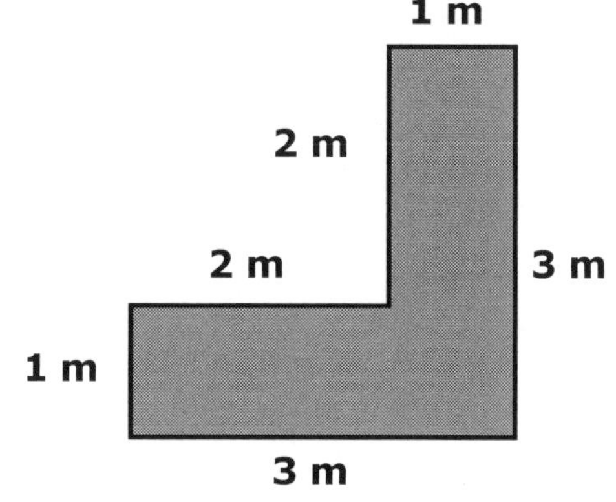

A. 10 m
B. 12 m
C. 16 m
D. 9 m

(4.MD.A.3)

CHAPTER REVIEW

9. What is the perimeter of the shape? ___________

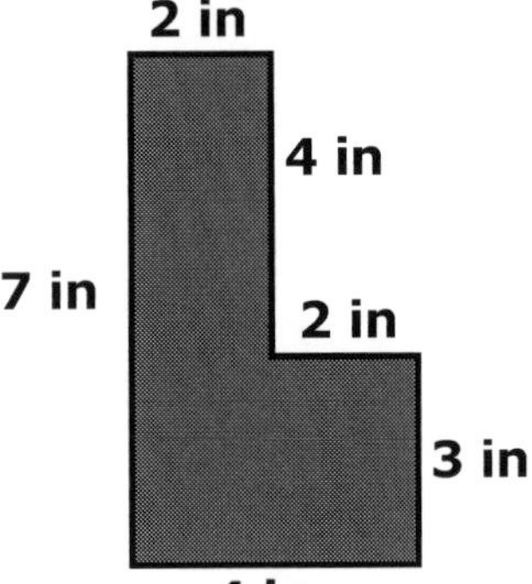

A. 20 in.
B. 22 in.
C. 18 in.
D. 28 in.

(4.MD.A.3)

10. This line plot shows the amount of fuel, in gallons, used for 7 campsite grills. How many gallons of fuel do most of the grills use?

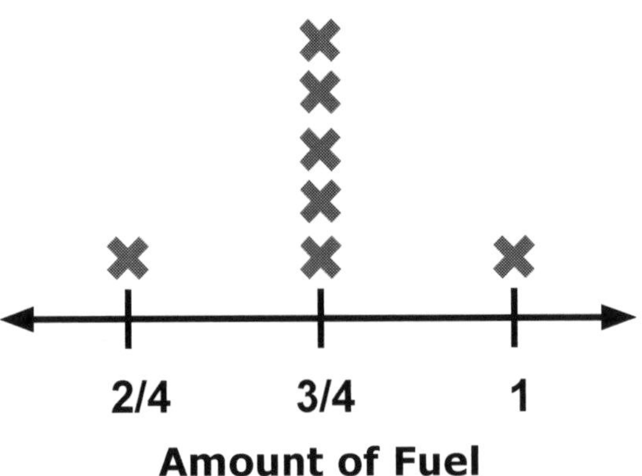

____

(4.MD.B.4)

11. This line plot shows the length of Mr. Green's pencils. How many pencils are longer than 8 ½ inches?

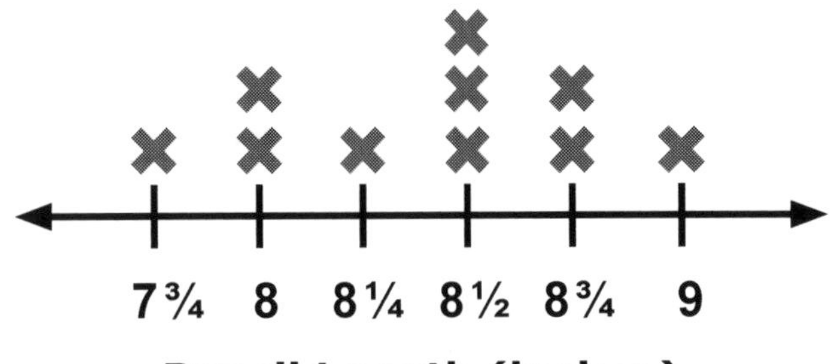

____

(4.MD.B.4)

12. What is the measure of this angle? Choose the best estimate.

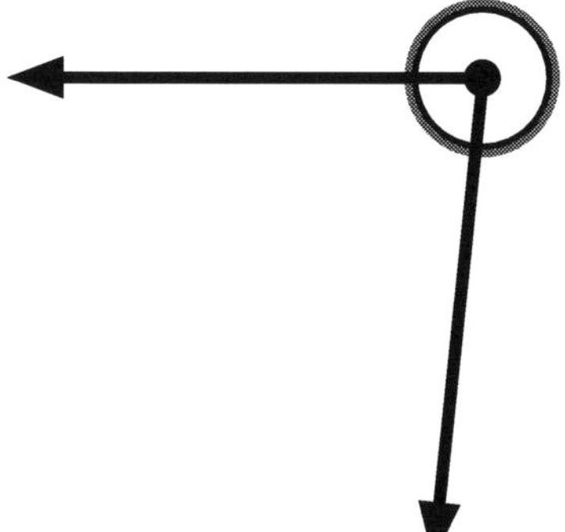

A. 1200

B. 900

C. 600

D. 850

(4.MD.C.5.A)

13. What fraction of a turn is this angle? ____________

(4.MD.C.5.A)

14. How many 1-degree angles are in an angle that turns through a full circle?

____

(4.MD.C.5.B)

15. How many 1-degree angles are in an angle that turns through a half-circle?

____

(4.MD.C.5.B)

CHAPTER REVIEW

16. Marlene draws an angle twice as large as the measure of the angle shown. What is the measure of Marlene's angle?

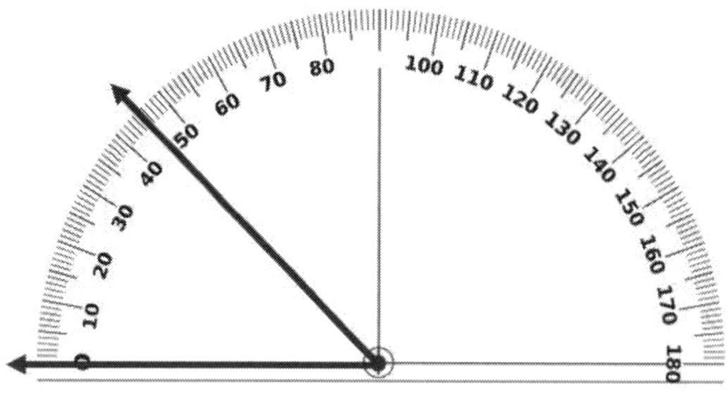

A. 45° **B.** 90°
C. 80° **D.** 27°

4.MD.C.6

17. Shyla believes the measure of this angle is 80 degrees. Do you agree with Shyla?

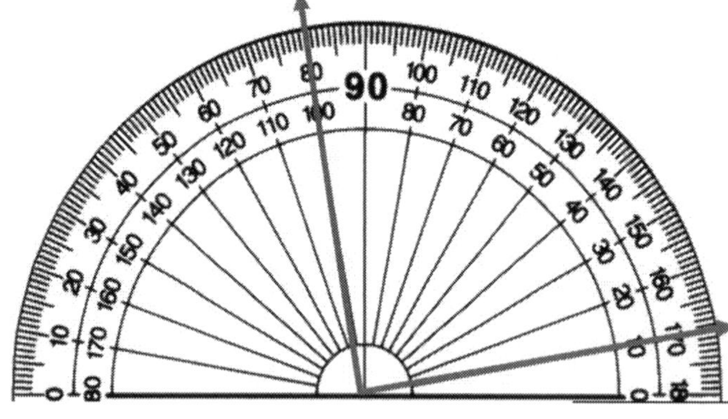

____

____

____

4.MD.C.6

18. What is the measure of Angle Y?

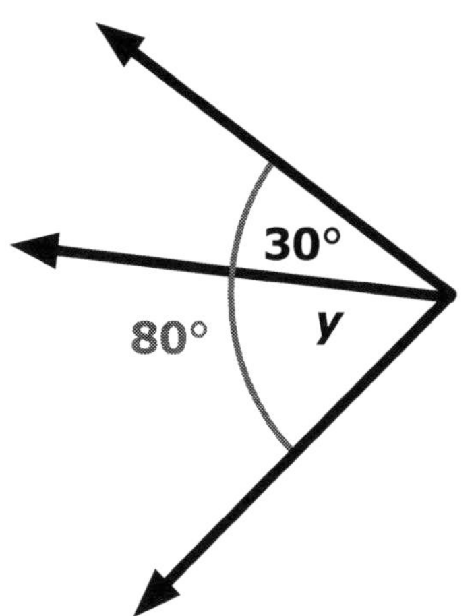

A. 40° **B.** 80°
C. 30° **D.** 50°

(4.MD.C.7)

19. What is the measure of Angle K?

70°
30°
k

A. 40° **B.** 50°
C. 30° **D.** 100°

(4.MD.C.7)

20. Mr. Wilson is building 2 new fences to separate his livestock. His property has the shape of a square.

Mr. Wilson's Farm

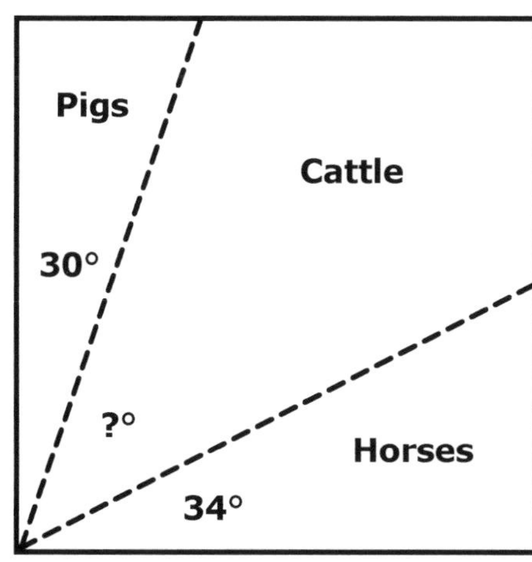

What is the measure, in degrees, of the missing angle?

____

(4.MD.C.7)

MEASUREMENT & DATA

EXTRA PRACTICE

1. Complete the table:

pints	3	4	☐	☐	8
cups	☐	☐	12	14	☐

(4.MD.A.1)

2. Which is more, 1,915 millimeters or 1 meter? ________________

(4.MD.A.1)

3. Which is more, 780 meters or 1 kilometer? ________________

(4.MD.A.1)

4. Madison's basketball practice started at 2:40 P.M. The team practiced offense for 1 hour and 10 minutes and defense for 1 hour and 45 minutes. What time did Madison's basketball practice end?

A. 5:35 P.M. **B.** 6:05 P.M. **C.** 5:15 P.M. **D.** 4:50 P.M.

(4.MD.A.2)

5. Xavier works as a teller at a bank downtown. Each morning, Xavier walks 10 minutes to the subway station. The subway ride to the bank takes 25 minutes. If Xavier works 5 days each week, how many hours does he spend traveling to his job at the bank?

A. 3 hours **B.** 2 hours and 55 minutes

C. 2 hours and 75 minutes **D.** 3 hours and 55 minutes

(4.MD.A.2)

6. Chippy got home from school at 1:15 P.M. and played video games for 1 hour. Then, it took him 1 hour and 35 minutes to finish his homework. What time did Chippy finish his homework?

A. 3:00 P.M. **B.** 3:50 P.M.

C. 3:15 P.M. **D.** 3:45 P.M.

4.MD.A.2

7. What is the perimeter of the shape?

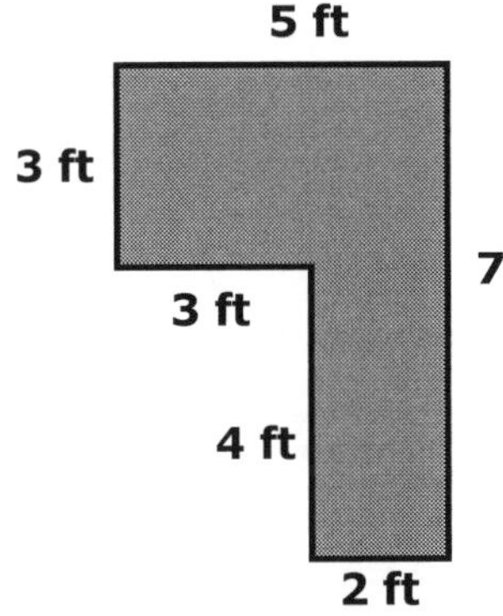

A. 28 ft **B.** 22 ft

C. 24 ft **D.** 35 ft

4.MD.A.3

8. The dimensions of a science notebook is 9 inches tall and 6 inches wide. What is its area?

A. 60 square inches **B.** 65 square inches

C. 58 square inches **D.** 54 square inches

4.MD.A.3

9. A square mirror has sides that are 5 feet long. What is the mirror's area?

A. 25 square feet **B.** 35 square feet

C. 15 square feet **D.** 20 square feet

4.MD.A.3

EXTRA PRACTICE

EXTRA PRACTICE

10. This line plot shows the amount of paint, in cups, 6 people used in art class. How many people used less than ¾ cups of white paint?

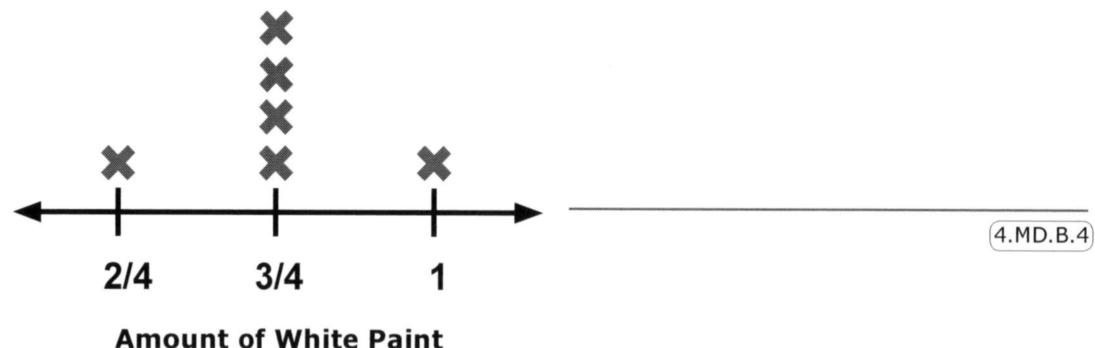

____

(4.MD.B.4)

11. This line plot shows the amount of paint, in cups, 6 people used in art class. How many people used less than 1 cup of white paint?

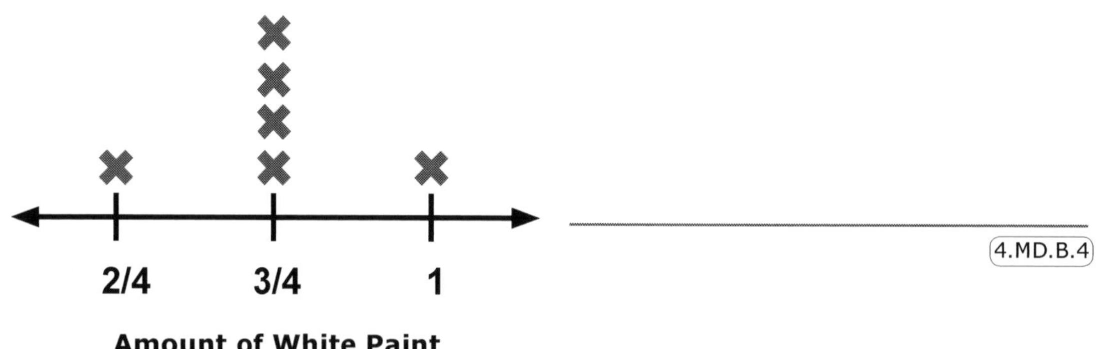

____

(4.MD.B.4)

12. The measure of angle H is 300. What fraction of a complete circle is the measure of angle H?

____

(4.MD.C.5.A)

EXTRA PRACTICE

13. Which fraction of a complete circle is the best estimate for the measure of this angle?

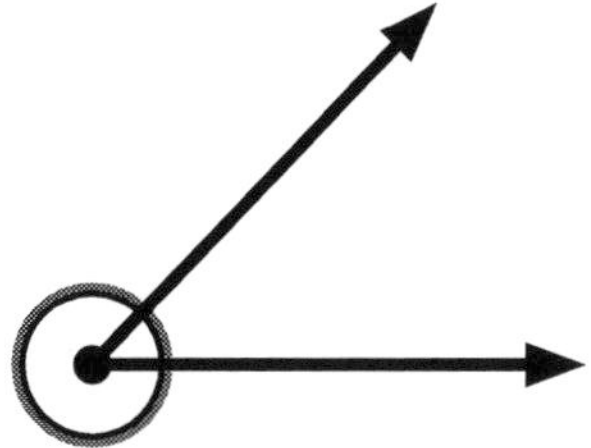

A. $\frac{360}{360}$ **B.** $\frac{60}{1}$ **C.** $\frac{45}{360}$ **D.** $\frac{360}{90}$

(4.MD.C.5.A)

14. Aziz draws an angle that turns through 15 one-degree angles. He copies the angle 2 more times to create a larger angle. What is the combined measure, in degrees, of the angles Aziz draws?

____

(4.MD.C.5.B)

15. The wheels on a bicycle turn 1 degree every $\frac{1}{360}$ seconds. How many degrees does the bicycle wheel turn in 2 seconds?

____

(4.MD.C.5.B)

EXTRA PRACTICE

16. Terry is drawing a 55 angle. Draw the second ray to create this angle.

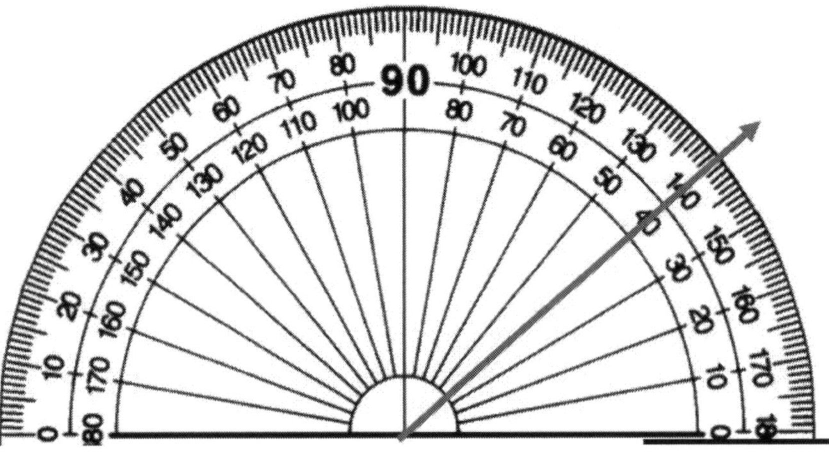

(4.MD.C.6)

17. Write 4 expressions that determine the measure of this angle.

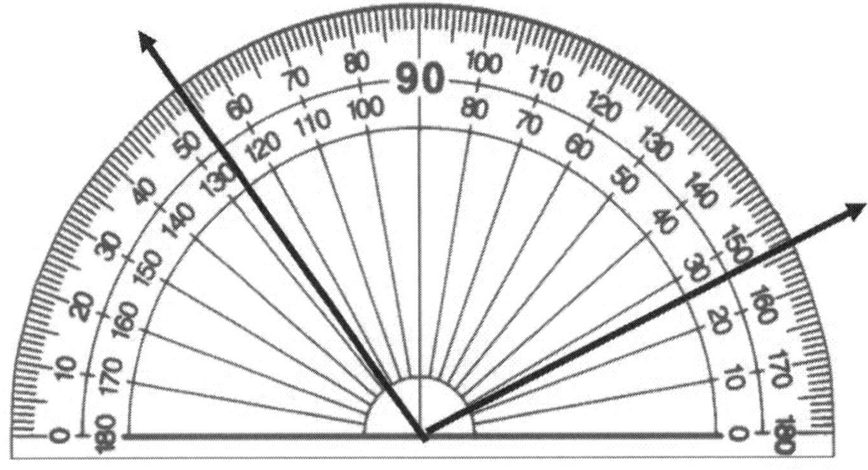

(4.MD.C.6)

EXTRA PRACTICE

18. The measure of angle WXY is 110. Kaufman states that the measure of Angle ZXY is 42. Do you agree with Kaufman?

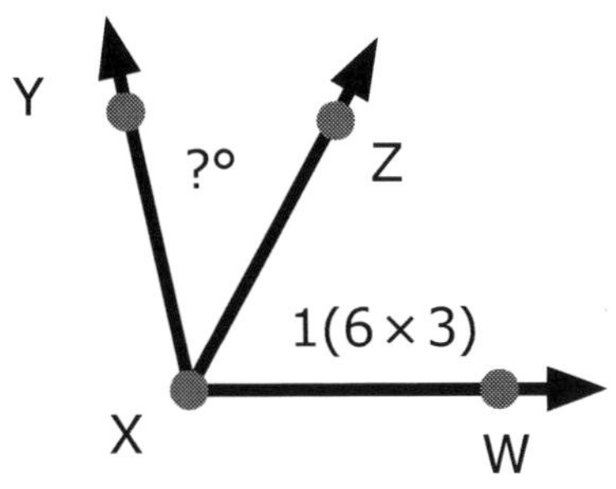

____

____

____

(4.MD.C.7)

19. Write the steps you would use to find the combined measure, in degrees, of these two angles.

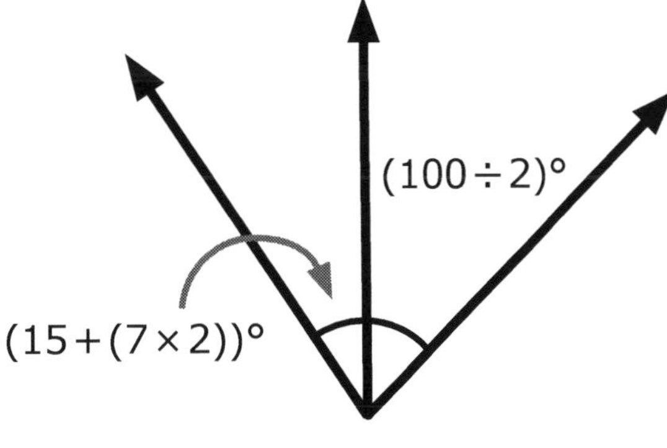

____

____

____

(4.MD.C.7)

EXTRA PRACTICE

20. Akiko is creating a geometric drawing using her protractor. She starts by drawing a hexagon and creating 2 triangles and a quadrilateral inside of it.

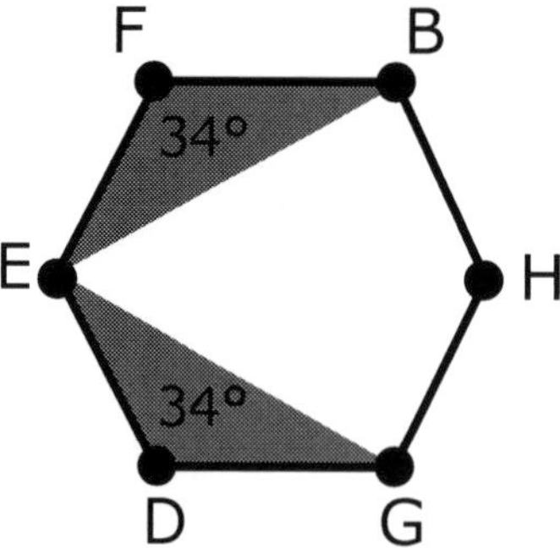

Angle DEF has a measure of 120°. What is the measure, in degrees, of Angle BEG?

____

(4.MD.C.7)

GEOMETRY

4

LINES AND ANGLES 143

TWO-DIMENSIONAL SHAPES 149

LINE OF SYMMETRY 154

CHAPTER REVIEW 158

EXTRA PRACTICE 162

1. Identify the following:

A. line
B. ray
C. line segment
D. edge

4.G.A.1

2. Identify the following:

A. line
B. ray
C. line segment
D. point

4.G.A.1

3. Identify the following:

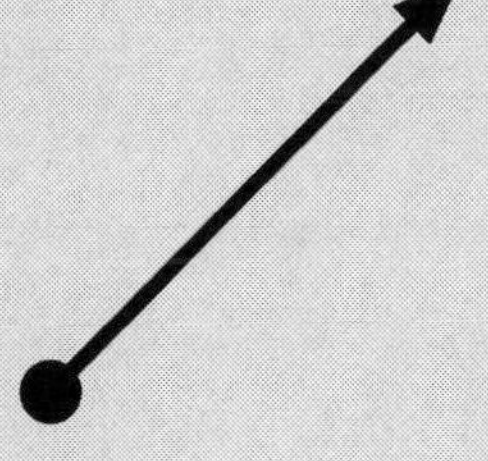

A. line
B. ray
C. line segment
D. angle

4.G.A.1

4. Identify the following:

A. line
B. ray
C. line segment
D. arrow

4.G.A.1

5. Identify the following:

A. line
B. ray
C. line segment
D. point

4.G.A.1

6. Millie draws this picture. What does she draw?

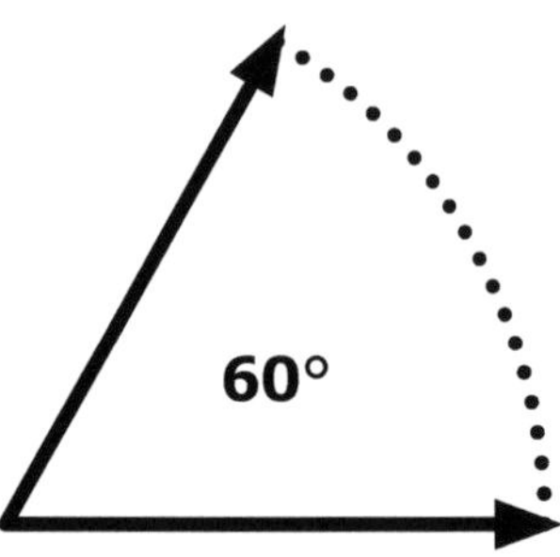

A. line
B. angle
C. line segment
D. ray

4.G.A.1

7. Andy connects Point X and Point Y on this circle. What does he use to connect the two points?

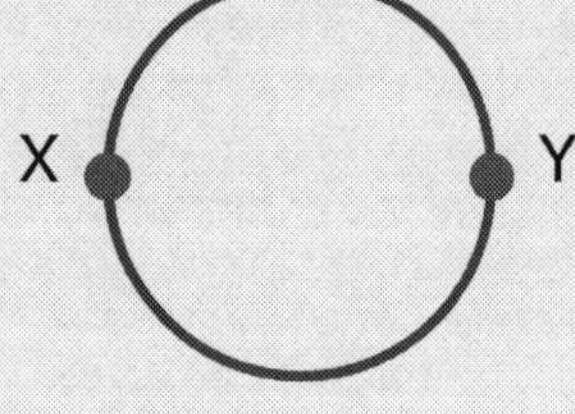

A. line
B. ray
C. line segment
D. slanting line

4.G.A.1

8. Identify the following:

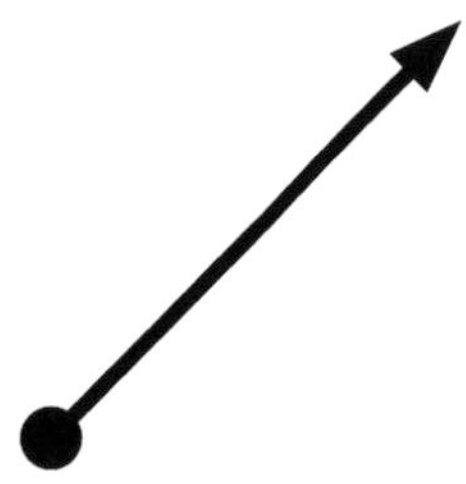

A. line
B. ray
C. line segment
D. edge

4.G.A.1

9. Identify the following:

A. line
B. ray
C. arrow
D. line segment

4.G.A.1

10. In the shape shown below, some line segments are parallel. Which of the following line segments are parallel?

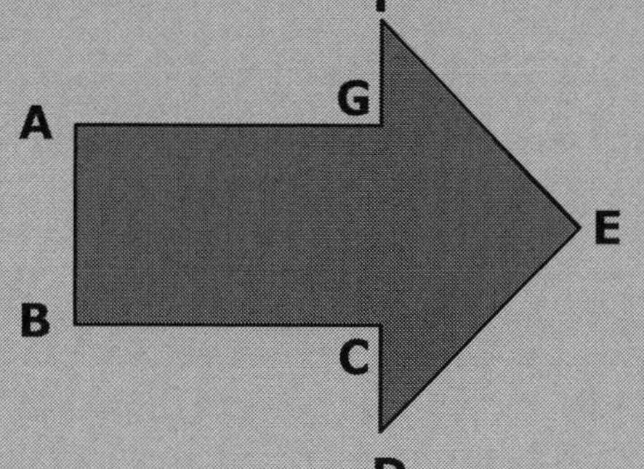

A. $\overline{FG}$ and $\overline{FE}$
B. $\overline{AB}$ and $\overline{ED}$
C. $\overline{AG}$ and $\overline{BC}$
D. $\overline{FE}$ and $\overline{DE}$

4.G.A.1

11. In the shape shown below, some line segments are perpendicular. Which of the following line segments are perpendicular?

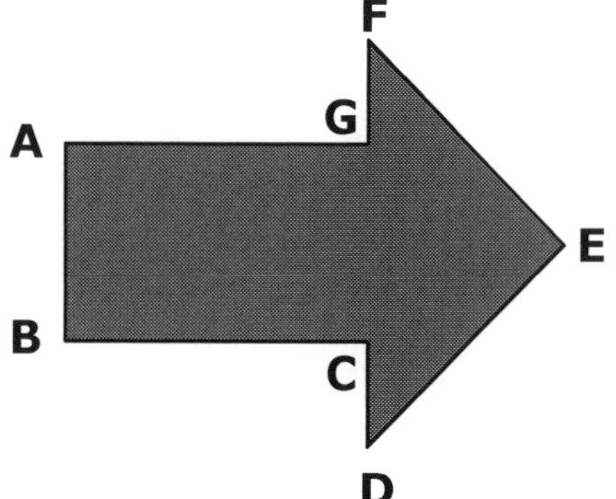

A. $\overline{AG}$ and $\overline{BC}$
B. $\overline{FE}$ and $\overline{DE}$
C. $\overline{FG}$ and $\overline{FE}$
D. $\overline{AB}$ and $\overline{ED}$

4.G.A.1

GEOMETRY

12. Which line segments are parallel in this shape?

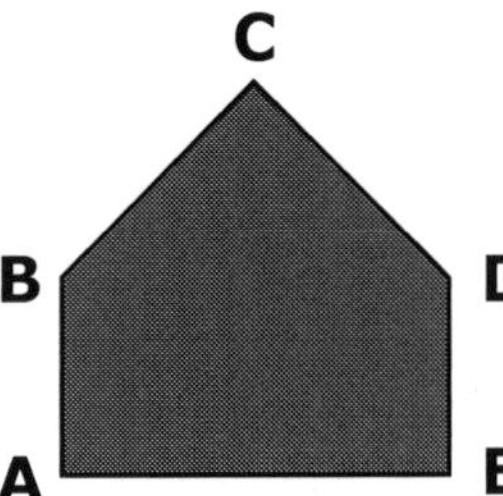

A. $\overline{AE}$ and $\overline{BC}$

B. $\overline{AB}$ and $\overline{ED}$

C. $\overline{BC}$ and $\overline{CD}$

D. $\overline{DE}$ and $\overline{BC}$

4.G.A.1

13. In the figure below, all angles that appear to be right angles are right angles. How many pairs of parallel line segments are in this shape?

A. 5 **B.** 4

C. 3 **D.** 2

4.G.A.1

14. True or False: This triangle has one right angle.

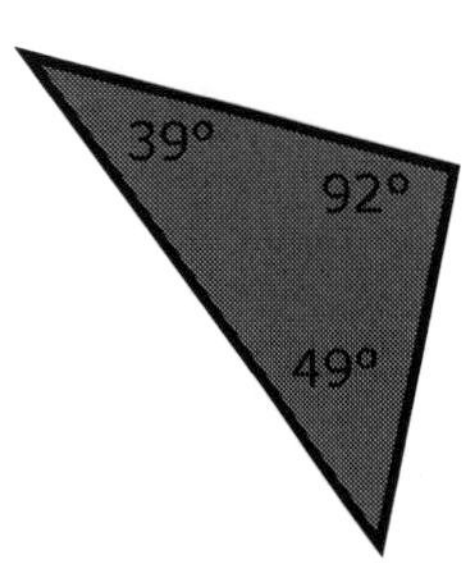

A. True **B.** False

4.G.A.1

15. True or False: This triangle has 3 acute angles.

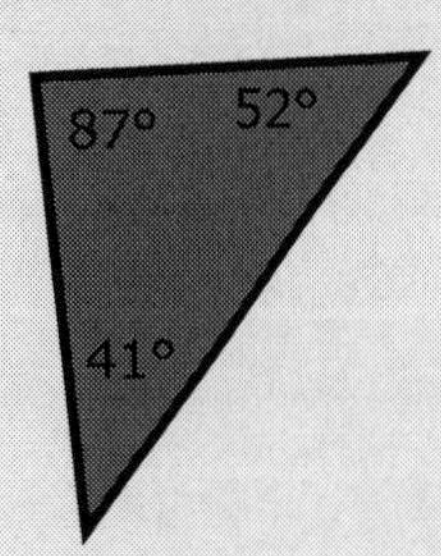

A. True **B.** False

4.G.A.1

16. True or False: This triangle has 1 obtuse angle.

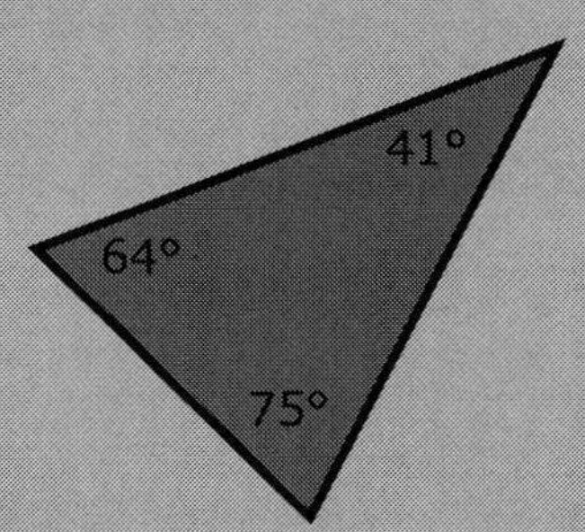

A. True **B.** False

4.G.A.1

17. True or False: This triangle has 1 obtuse angle.

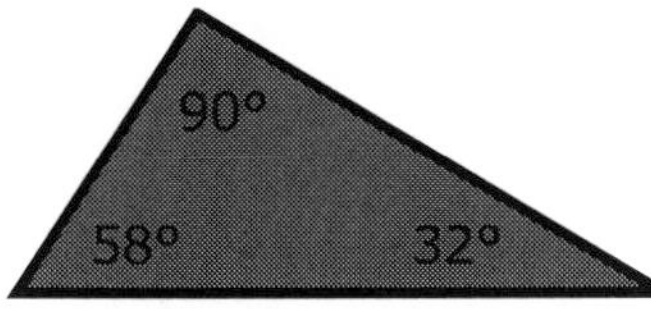

A. True **B.** False

4.G.A.1

18. True or False: Two pairs of opposite sides are parallel.

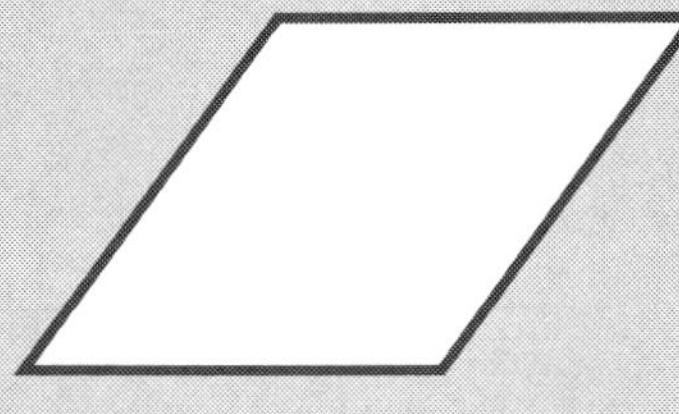

A. True **B.** False

4.G.A.1

19. True or False: 2 pairs of opposite sides are parallel.

A. True **B.** False

4.G.A.1

20. True or False: These lines are perpendicular.

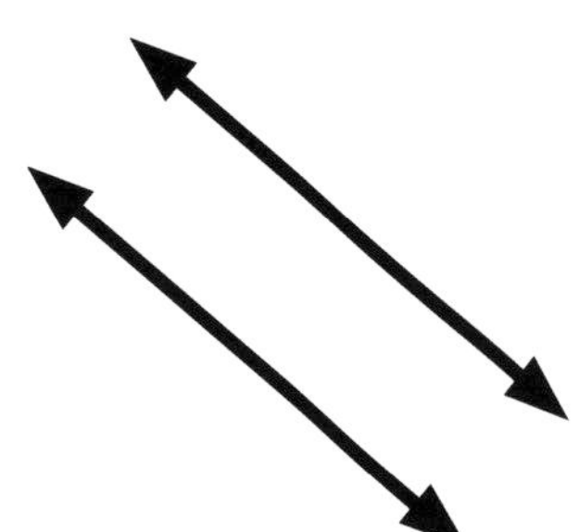

A. True **B.** False

4.G.A.1

GEOMETRY

1. What name best describes this shape?

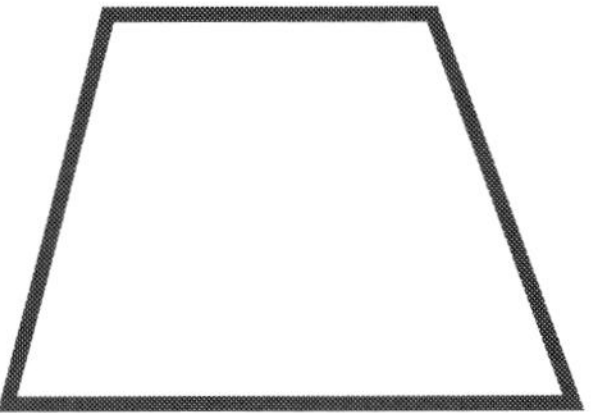

A. square
B. kite
C. quadrilateral
D. rectangle

4.G.A.2

2. How many right angles does this quadrilateral have?

A. 2
B. 3
C. 6
D. 4

4.G.A.2

3. What name best describes this shape?

A. parallelogram
B. trapezoid
C. rhombus
D. square

4.G.A.2

4. What name best describes this shape?

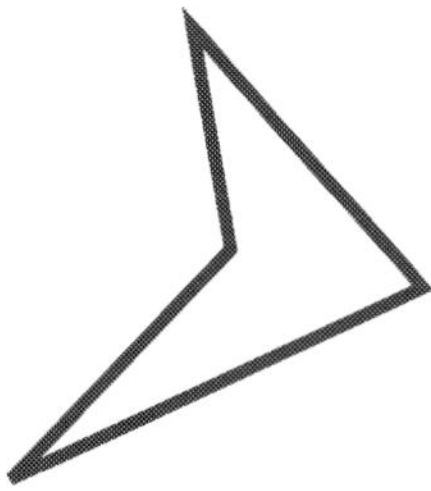

A. quadrilateral
B. parallelogram
C. trapezoid
D. kite

4.G.A.2

NAME: .... DATE: ....

GEOMETRY

5. What name best describes this shape?

A. square
B. rhombus
C. rectangle
D. parallelogram

4.G.A.2

6. What name best describes this shape?

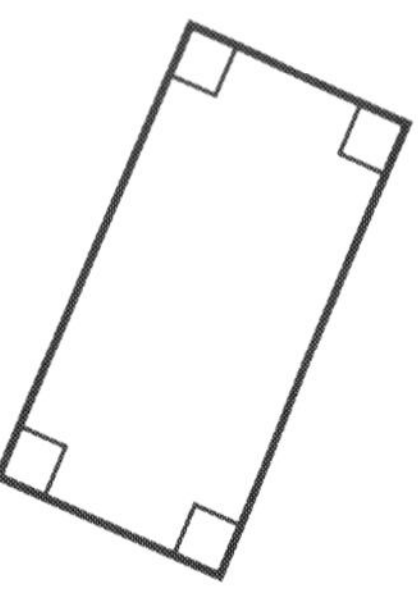

A. square
B. rhombus
C. rectangle
D. kite

4.G.A.2

7. What name best describes this shape?

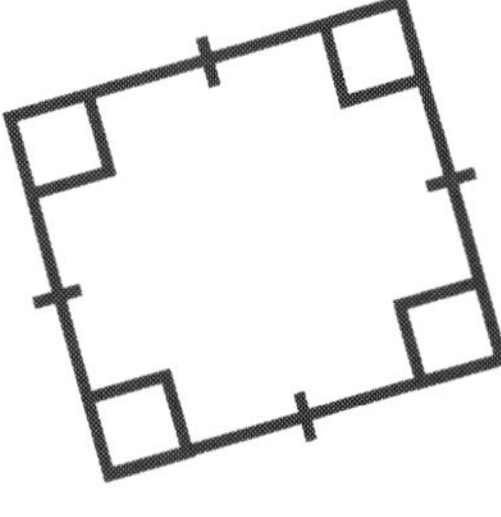

A. square
B. rhombus
C. rectangle
D. pentagon

4.G.A.2

8. What name best describes this shape?

A. rhombus
B. trapezoid
C. pentagon
D. quadrilateral

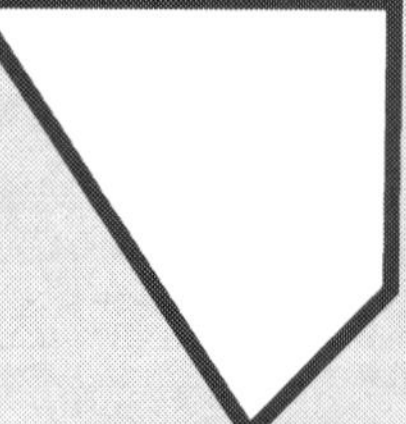

4.G.A.2

9. What name best describes this shape?

A. rhombus
B. trapezoid
C. quadrilateral
D. rectangle

4.G.A.2

10. What type of triangle is shown?

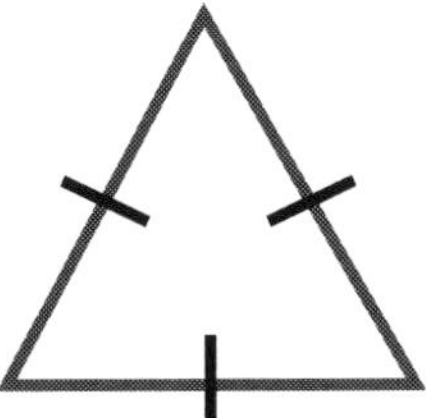

A. equilateral
B. right
C. obtuse
D. scalene

4.G.A.2

11. What type of triangle is shown?

A. right
B. isosceles
C. scalene
D. equilateral

4.G.A.2

12. What type of triangle is shown?

A. scalene
B. equilateral
C. right
D. obtuse

4.G.A.2

TWO-DIMENSIONAL SHAPES

13. What type of triangle is shown?

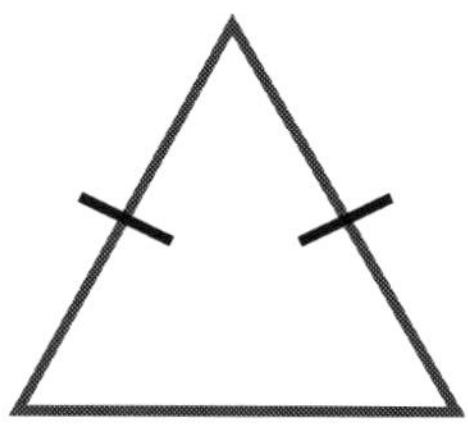

A. isosceles
B. equilateral
C. right
D. obtuse

4.G.A.2

14. Which shapes is a trapezoid?

A.  **B.**  **C.**  **D.** 

4.G.A.2

15. Which shape is a rectangle?

A. 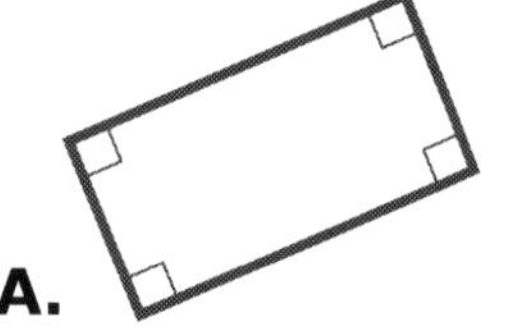 **B.** 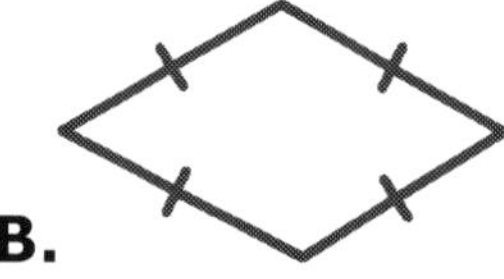 **C.** 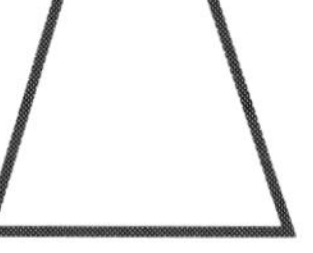 **D.** 

4.G.A.2

16. Which shape is a rhombus?

A.  **B.** 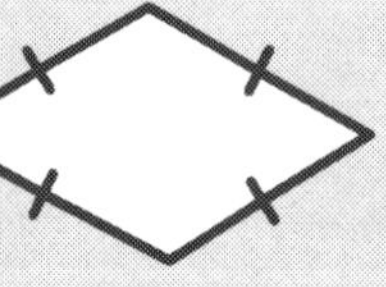 **C.**  **D.** 

4.G.A.2

17. Which shape has 1 pair of perpendicular line segments?

A.  **B.**  **C.** 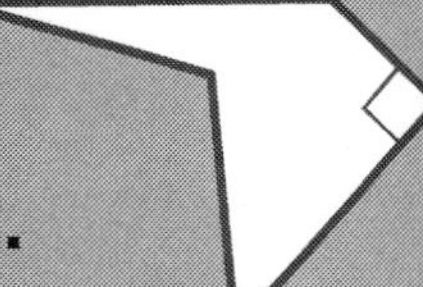 **D.** 

4.G.A.2

18. True or False: The name that best describes this shape is a quadrilateral.

A. True **B.** False

4.G.A.2

19. Amanda drew a shape with 4 sides and 4 right angles. Which is the best description of the shape she drew?

A. rectangle **B.** rhombus **C.** kite **D.** hexagon

4.G.A.2

20. True or False: The name that best describes this shape is the quadrilateral.

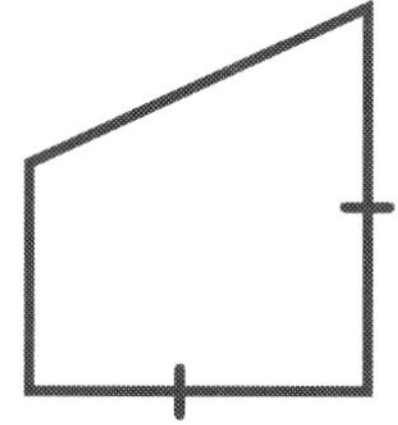

A. True

B. False

4.G.A.2

GEOMETRY

1. **True or False:** The dotted line in the image below is a line of symmetry.

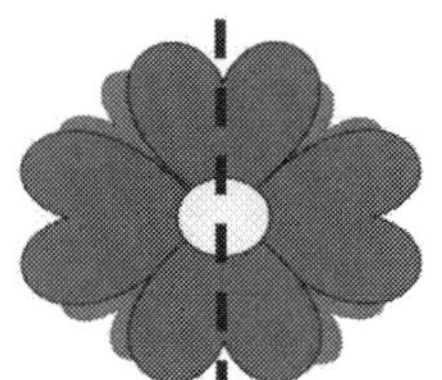

A. True
B. False

(4.G.A.3)

2. **True or False:** The dotted line in the image below is a line of symmetry.

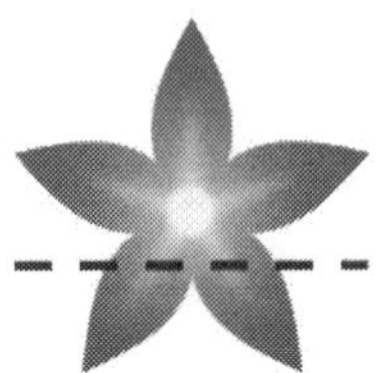

A. True
B. False

(4.G.A.3)

3. **True or False:** The dotted line in the image below is a line of symmetry.

A. True
B. False

(4.G.A.3)

4. **True or False:** The dotted line in the image below is a line of symmetry.

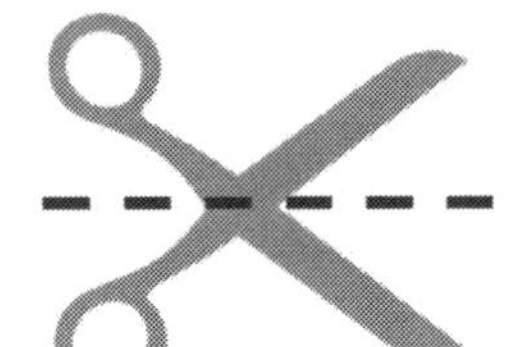

A. True
B. False

(4.G.A.3)

5. **True or False:** The dotted line in the image below is a line of symmetry.

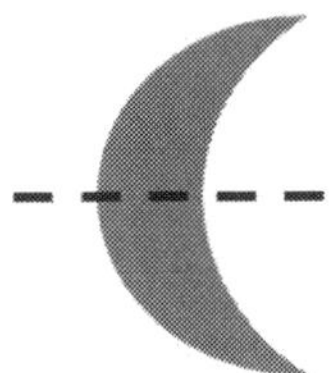

A. True
B. False

(4.G.A.3)

6. **True or False:** The dotted line in the image below is a line of symmetry.

A. True
B. False

(4.G.A.3)

7. **True or False:** The dotted line in the image below is a line of symmetry.

A. True
B. False

(4.G.A.3)

8. **True or False:** The dotted line in the image below is a line of symmetry.

A. True
B. False

4.G.A.3

9. **True or False:** The dotted line in the image below is a line of symmetry.

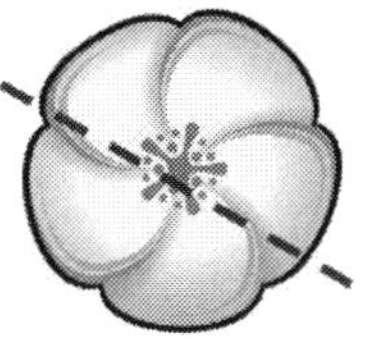

A. True
B. False

4.G.A.3

10. **True or False:** The dotted line in the image below is a line of symmetry.

A. True
B. False

4.G.A.3

11. **True or False:** The capital letter "H" has 3 lines of

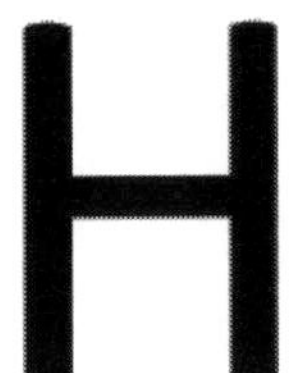

A. True
B. False

4.G.A.3

12. **True or False:** The letter "C" has 1 line of symmetry.

A. True
B. False

4.G.A.3

13. **True or False:** The dotted line in the image below is a line of symmetry.

A. True
B. False

4.G.A.3

14. **True or False:** The image of the ladybug has 1 line of symmetry.

A. True
B. False

4.G.A.3

15. **True or False:** The image has only 1 line of symmetry.

A. True
B. False

4.G.A.3

16. True or False: The picture of the giraffe has 1 line of symmetry.

A. True
B. False

(4.G.A.3)

17. True or False: The image below has 0 lines of symmetry.

A. True
B. False

(4.G.A.3)

18. Draw the lines of symmetry on the image below.

(4.G.A.3)

19. Draw the lines of symmetry on the image below.

(4.G.A.3)

20. Draw the lines of symmetry on the shape below

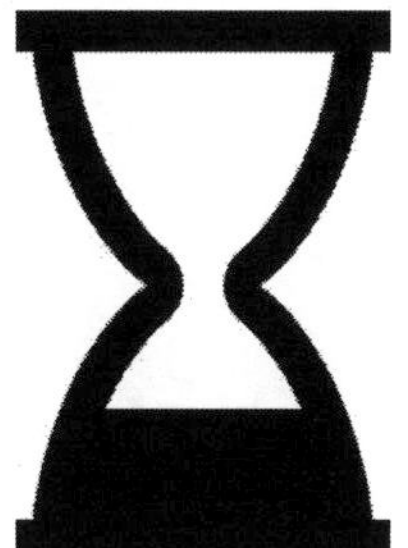

(4.G.A.3)

NAME: .... DATE: ....

GEOMETRY

CHAPTER REVIEW

1. What types of angles are in this triangle? ____________

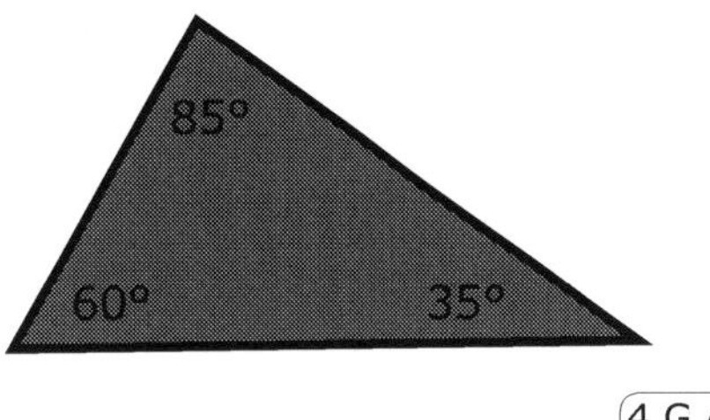

4.G.A.1

2. What type of triangle is this triangle? ____________

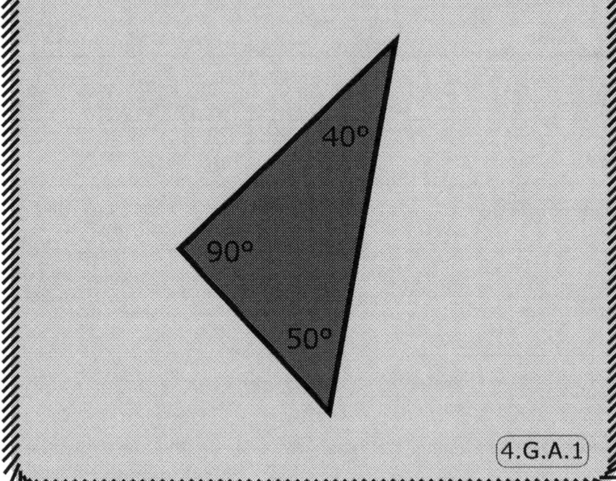

4.G.A.1

3. What type of angle is shown? ____________

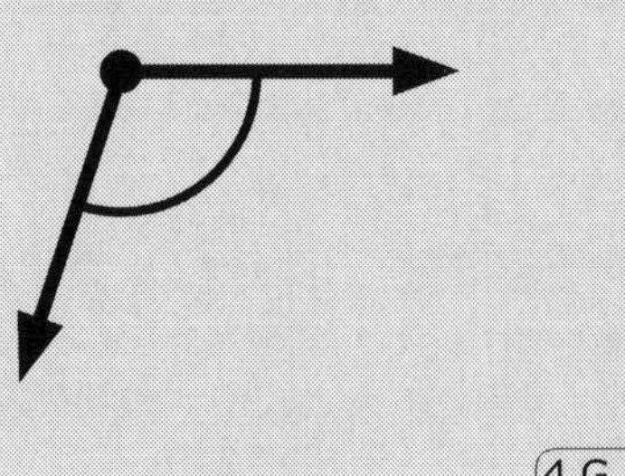

4.G.A.1

4. What type of angle is shown? ____________

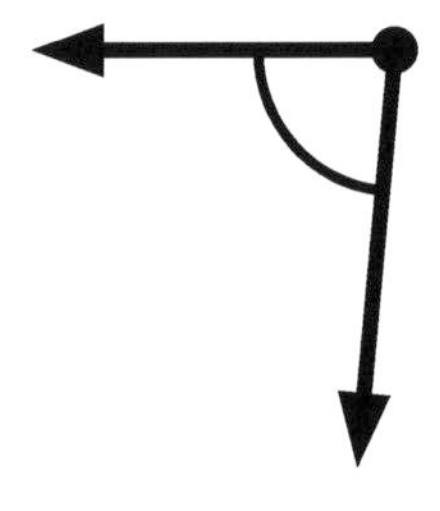

4.G.A.1

5. What type of angle is shown? ____________

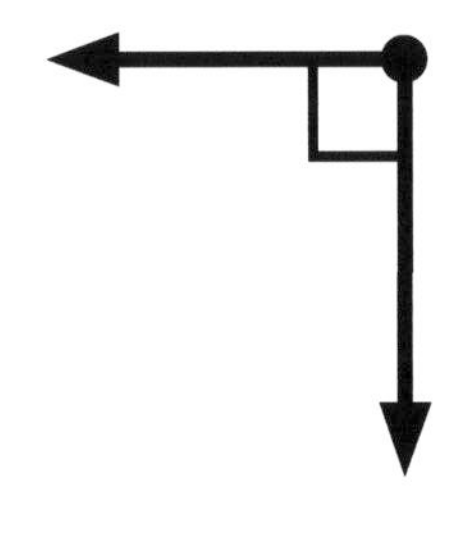

4.G.A.1

6. What type of angle is shown? ____________

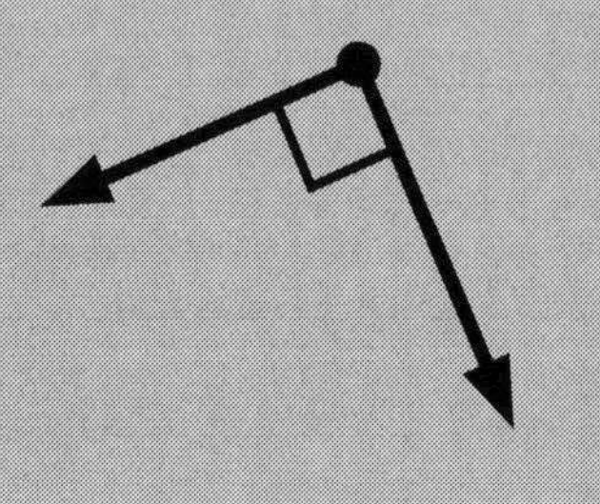

4.G.A.1

7. What type of angle is shown? ____________

(4.G.A.1)

8. **True or False:** The name that best describes this shape is the rhombus.

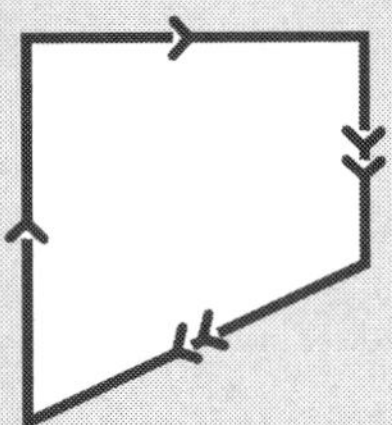

A. True **B.** False

(4.G.A.2)

9. **True or False:** The triangle below is a right triangle.

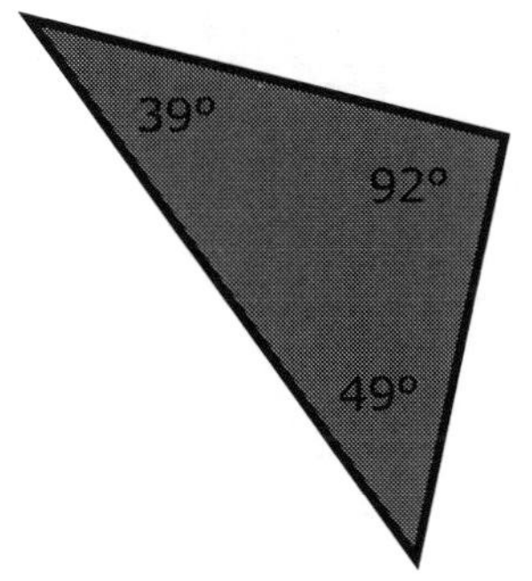

A. True **B.** False

(4.G.A.2)

10. **True or False:** The following triangle is an acute triangle.

87° 52°
41°

A. True **B.** False

(4.G.A.2)

11. **True or False:** This quadrilateral is a trapezoid.

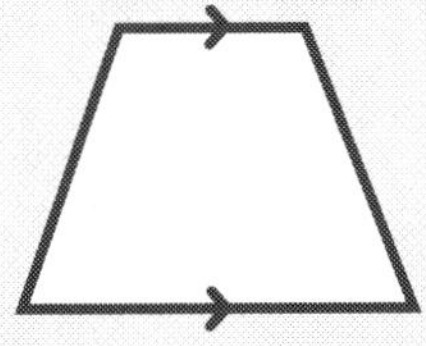

A. True **B.** False

(4.G.A.2)

CHAPTER REVIEW

12. True or False: This quadrilateral is a trapezoid.

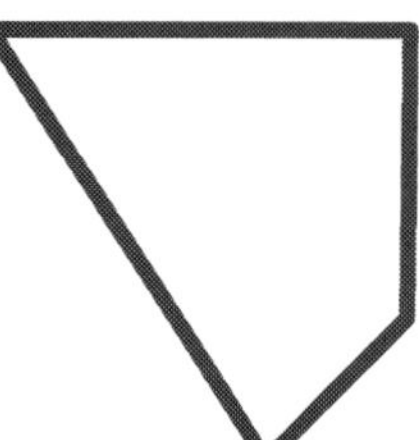

A. True **B.** False

(4.G.A.2)

13. True or False: This quadrilateral is a trapezoid.

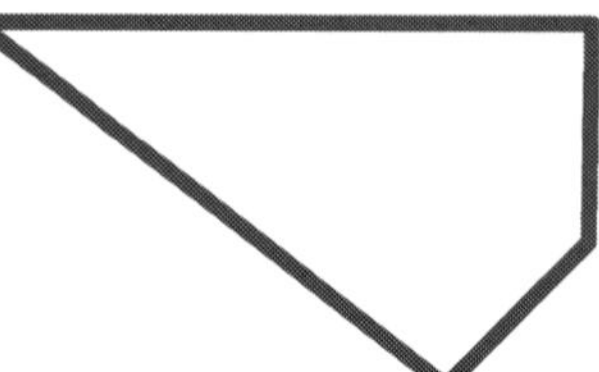

A. True **B.** False

(4.G.A.2)

14. True or False: This quadrilateral is a trapezoid.

A. True **B.** False

(4.G.A.2)

15. How many lines of symmetry does this shape have?

A. 1 **B.** 2
C. 3 **D.** 4

(4.G.A.3)

16. How many lines of symmetry does this shape have?

A. 2 **B.** 1
C. 3 **D.** 0

(4.G.A.3)

17. How many lines of symmetry does this shape have?

A. 1 **B.** 2
C. 3 **D.** 4

4.G.A.3

18. How many lines of symmetry does this shape have?

A. 4 **B.** 2
C. 1 **D.** 0

4.G.A.3

19. How many lines of symmetry does this shape have?

A. 1 **B.** 2
C. 3 **D.** 4

4.G.A.3

20. True or False. This image has exactly 2 lines of symmetry.

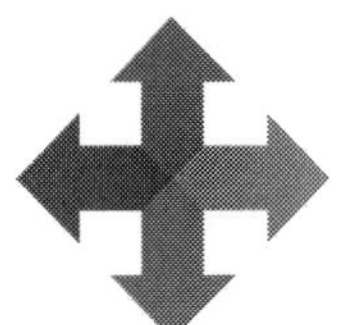

____

4.G.A.3

NAME: .... DATE: ....

GEOMETRY

EXTRA PRACTICE

1. How many pairs of rays in this diagram are perpendicular?

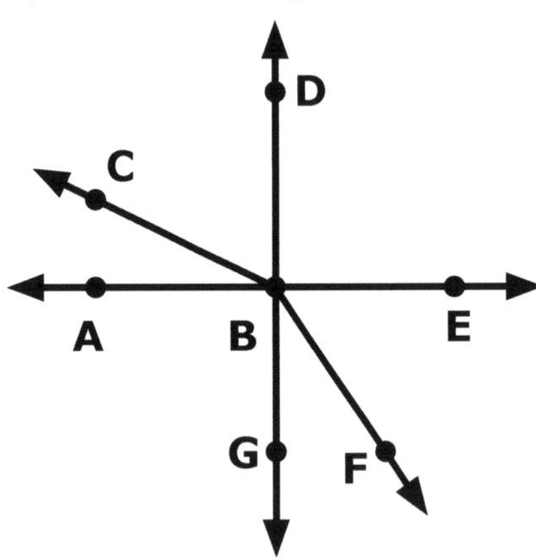

____

4.G.A.1

2. Adam draws a polygon. All of the interior angles are obtuse angles. Which shape did he draw?

A.  **B.** 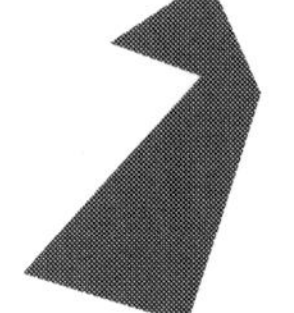 **C.**  **D.** 

4.G.A.1

3. Which letters can be used to describe an angle shown in this diagram?

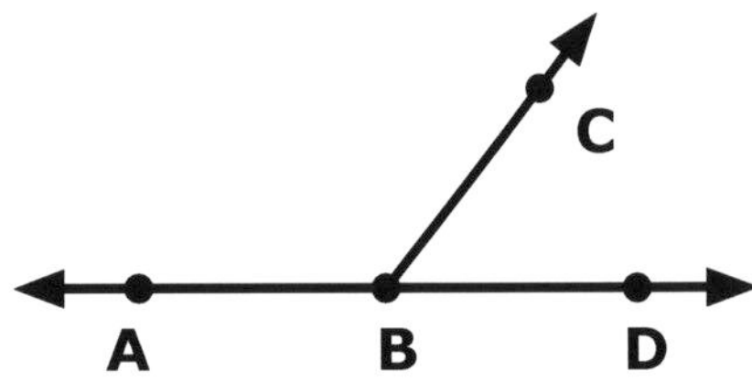

A. B **B.** ABC **C.** BCD **D.** BAD

4.G.A.1

4. How many parallel lines exist in this polygon?

____

(4.G.A.1)

5. A vertex of a polygon is the point where two edges meet. How many vertices does this polygon have?

____

(4.G.A.1)

6. List the three different two-dimensional shapes do you see in this picture.

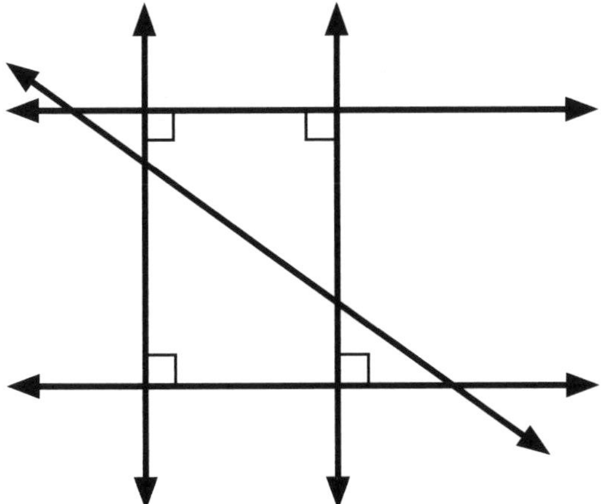

____

____

____

(4.G.A.1)

7. Heni creates a square on this geoboard using a rubber band.

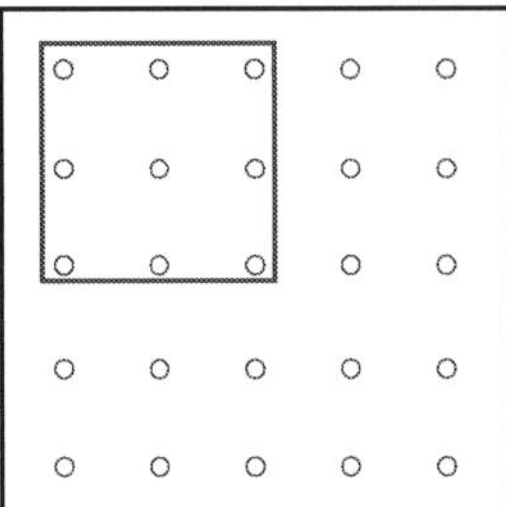

How could she move one of the corners to create a pentagon?

____

____

____

(4.G.A.1)

8. Which statement correctly describes the shapes in each group?

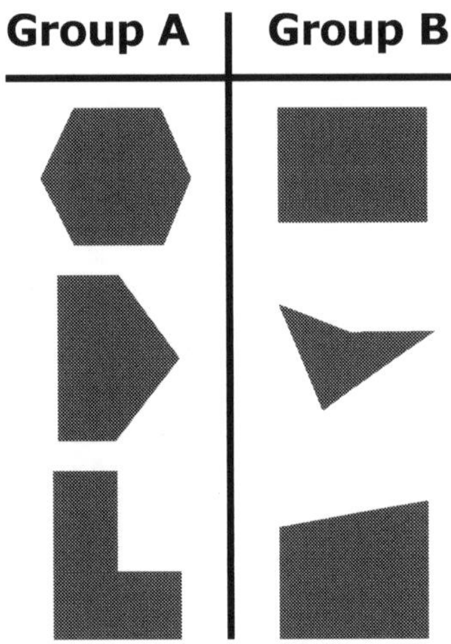

A. The shapes in Group A are all hexagons.

B. The shapes in Group B are all quadrilaterals.

C. The shapes in Group A have perpendicular line segments.

D. The shapes in Group B have parallel line segments.

(4.G.A.2)

GEOMETRY

9. Jamila says that all right triangles have two acute angles. Is she correct? Explain your reasoning.

____

____

____

4.G.A.2

10. Leanne describes this triangle as an equilateral triangle. Josh describes this triangle as an isosceles triangle. Can they both be correct? Explain your reasoning.

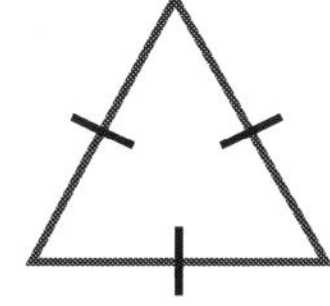

____

____

____

4.G.A.2

11. Which letter is in the shape of an octagon?

A. L **B.** V **C.** T **D.** E

4.G.A.2

EXTRA PRACTICE

12. Draw a shape with these characteristics:

- The shape has 4 sides.
- The shape is a parallelogram.
- The shape has opposite equal acute angles and opposite equal obtuse angles.

4.G.A.2

13. Is it possible to create a hexagon with 2 right angles? Explain your reasoning.

____

____

____

____

____

4.G.A.2

14. Which of the following states border a state that is shaped like a hexagon?

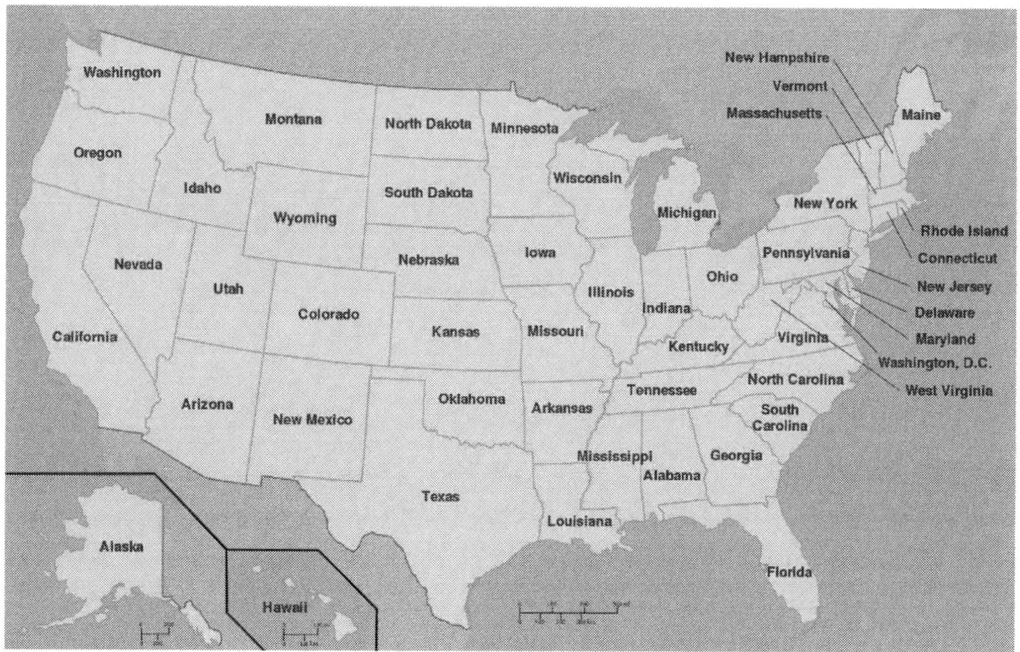

A. Arkansas and Mississippi **B.** Nevada and Oregon
C. Wyoming and Colorado **D.** Michigan and Indiana

4.G.A.2

15. Draw a shape or figure with exactly 3 lines of symmetry.

4.G.A.3

NAME: .... DATE: ....

GEOMETRY

16. Jocelynn believes the more sides a shape has, the more lines of symmetry it must have. Draw an example of when this is not true.

4.G.A.3

17. Lamar draws the following line of symmetry on this leaf. Do you agree with Lamar? Explain your reasoning.

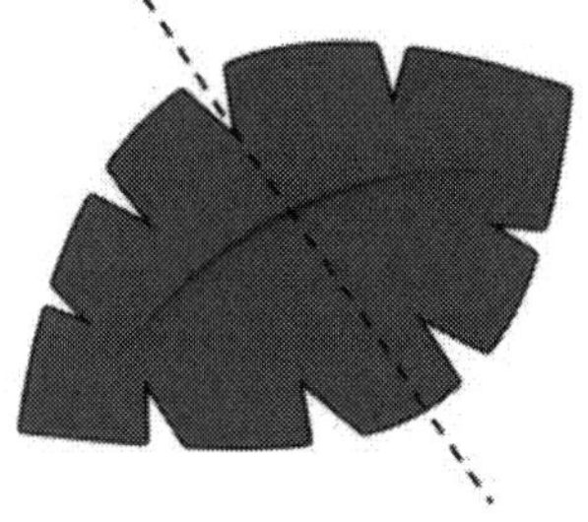

____

____

____

4.G.A.3

GEOMETRY

18. Which model is made up of 3 rectangles and has the greatest number of lines of symmetry?

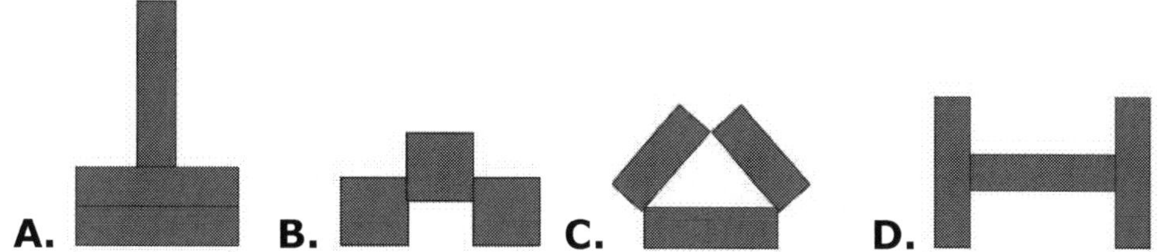

4.G.A.3

19. Leo says all quadrilaterals have lines of symmetry. Do you agree with Leo? Draw an example (or examples) to explain your reasoning.

4.G.A.3

20. What two-dimensional figure has more than an infinite number of lines of symmetry?

____

4.G.A.3

Ace Academic Publishing
ACHIEVING EXCELLENCE TOGETHER

COMPREHENSIVE ASSESSMENTS

4

Ace Academic Publishing
ACHIEVING EXCELLENCE TOGETHER

www.aceacademicprep.com

ASSESSMENT ①

1. Jack says that 5 × 6 means the same thing as having 5 and then get 6 more. Do you agree or disagree? Explain your reasoning.

____

____

____

4.OA.A.1

2. There is a rack of basketballs in the gym. The rack has 6 rows, and each row has 7 basketballs on it. There are 40 students in the gym. Are there enough basketballs for every student to have one? Explain how you know.

____

____

____

4.OA.A.2

3. James has 7 model cars. Michelle has twice as many model cars as James. How many more model cars do they need to have 40 model cars together?

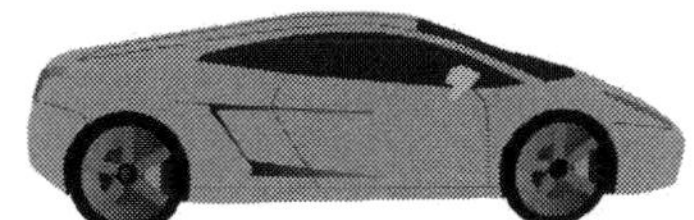

A. 47
B. 33
C. 21
D. 19

4.OA.A.3

4. Jane is 8 years old. She is 2 times older than her 4-year-old sister. Write an equation that matches this story. Explain your reasoning.

____

____

____

(4.OA.A.1)

5. Rachel says that no prime numbers are even other than the number 2. Do you agree or disagree? Why?

____

____

____

(4.OA.B.4)

6. Your little brother made this pattern fish in art class. What patterns did he use?

____

____

____

(4.OA.C.5)

7. A bus picked up 28 children at its second stop. This number is 7 times the number of children than it picked up at its first stop. How many children were picked up at the first stop?

A. 4 **B.** 5 **C.** 7 **D.** 35

(4.OA.A.2)

8. Diego scored 9 points in his first basketball game. He scored the same number of points in the next 20 games.

Which point on this number line best represents the total number of points Diego scored in all 21 games?

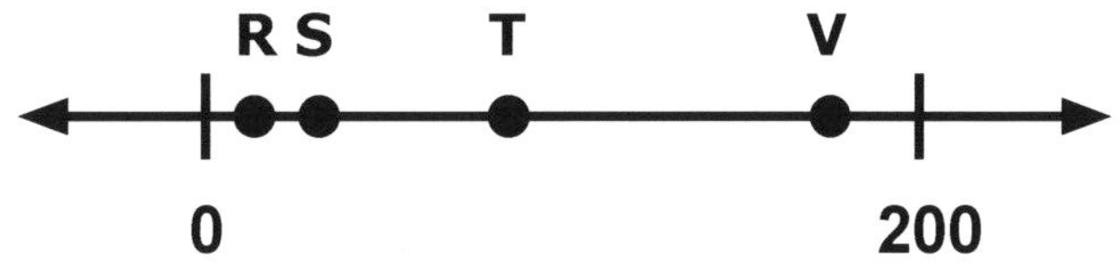

A. Point R **B.** Point S **C.** Point T **D.** Point V

(4.NBT.A.1)

9. The population of City Ville is 30,000 people. The population of Townson is 10 times the population of City Ville. What is the population of Townson?

____

(4.NBT.A.1)

10. There are 11 tables in the school lunchroom. Each table can seat 12 students. Today, 124 students are eating lunch. How many empty seats are there?

____

(4.OA.A.3)

ASSESSMENT 1

11. Sixty-four thousand seventy-two people attend the football game on Sunday. One-fourth of the people who attend the game are under the age of 18.

How many people who attend the game are over the age of 18?

A. Sixteen thousand eighteen
B. Forty-eight thousand fifty-four
C. Fifty-two thousand sixty-six
D. Forty-eight thousand five hundred forty

(4.NBT.A.2)

12. The population of Switzerland is 8,476,005.

How would this number be expressed in expanded form?

____

(4.NBT.A.2)

13. Veronica owns four different instruments.

- A piano valued at $5,145
- A clarinet valued at $2,014
- A saxophone valued at $1,965
- A violin valued at $4,308

If she rounds each value to the nearest thousand, what is the approximate value of Veronica's instruments?

A. $12,000 **B.** $13,400 **C.** $13,000 **D.** $15,000

(4.NBT.A.3)

14. A 4-digit number has the digits 1,4,9, and 8. When rounded to the nearest thousand, the number is labeled as Point A on this number line.

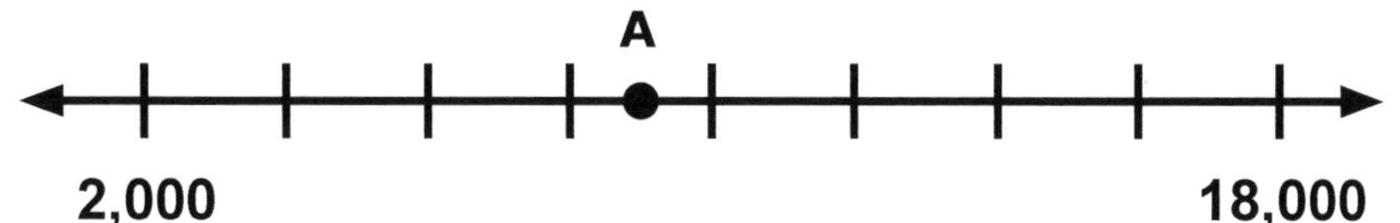

Write the number. ______________________

(4.NBT.A.3)

15. Violet uses the standard algorithm to subtract these numbers.

$$\begin{array}{r} 7{,}98\overset{11}{\not1} \\ -\ 5{,}5\overset{5}{\not4}8 \\ \hline 2{,}433 \end{array}$$

Do you agree with Violet's strategy? Explain your reasoning.

____

____

____

(4.NBT.B.4)

16. A truck driver travels 1,118 miles from Minneapolis, Minnesota to Dallas, Texas. He makes 4 round trips on this route.

How many miles does he travel?

A. 8,944 miles **B.** 4,472 miles
C. 1,122 miles **D.** 4,442 miles

(4.NBT.B.5)

ASSESSMENT 1

17. On Monday, an airplane flies 2,789 miles from New York City, New York to Los Angeles, California. The airplane then flies back to New York City on Tuesday.

If the airplane flies to Los Angeles on Wednesday, and back to New York on Thursday, what multiplication equation could you use to find the total number of miles traveled Monday through Thursday?

How many miles does the airplane travel from Monday through Thursday?

____

4.NBT.B.5

18. Angela travels 441 miles in 7 hours driving at a constant speed.

What is her driving speed?

A. 59 miles per hour
B. 62 miles per hour
C. 63 miles per hour
D. 60 miles per hour

4.NBT.B.6

19. A number divided by 8 has a quotient of 213. What is the number?

____

4.NBT.B.6

20. Jenny has 24 pairs of socks. Four of the pairs of socks are white and the rest are black. Choose the pair of fractions that shows equivalent fractions for the number of black socks in Jenny's collection.

A. $\frac{6}{14}$ and $\frac{3}{6}$
B. $\frac{8}{20}$ and $\frac{4}{10}$
C. $\frac{10}{12}$ and $\frac{5}{6}$
D. $\frac{8}{10}$ and $\frac{10}{12}$

4.NF.A.1

21. Sari and Bella collect stamps.

- $\frac{6}{8}$ of Sari's stamps are from California.
- $\frac{4}{5}$ of Bella's stamps are from California.

Which expression correctly compares the portion of Sari and Bella's stamps that are from California?

A. $\frac{6}{8} > \frac{4}{5}$ **B.** $\frac{6}{8} < \frac{4}{5}$ **C.** $\frac{6}{8} = \frac{4}{5}$ **D.** $\frac{3}{4} > \frac{4}{5}$

4.NF.A.2

22. This table shows the ingredients needed to make pizza dough.

Ingredient	flour	water	seasoning
Amount	$\frac{7}{8}$ cups	$\frac{2}{8}$ cups	$\frac{1}{8}$ cups

How many more cups of flour than water is used to make pizza dough?

____

4.NF.B.3

23. A recipe requires $\frac{3}{5}$ cups of flour and $\frac{1}{5}$ cups of sugar to make 4 servings.

Beth wants to double the recipe. How much flour and sugar will she need?

____

4.NF.B.4

COMPREHENSIVE ASSESSMENTS

24. The thickness of a penny is $1\frac{5}{10}$ millimeters. A nickel is $\frac{40}{100}$ millimeters thicker than a penny. What is the thickness, in mm, of a nickel?

____

4.NF.C.5

25. Aditya determines the distance between Point A and Point C on the number line below is $\frac{60}{100}$.

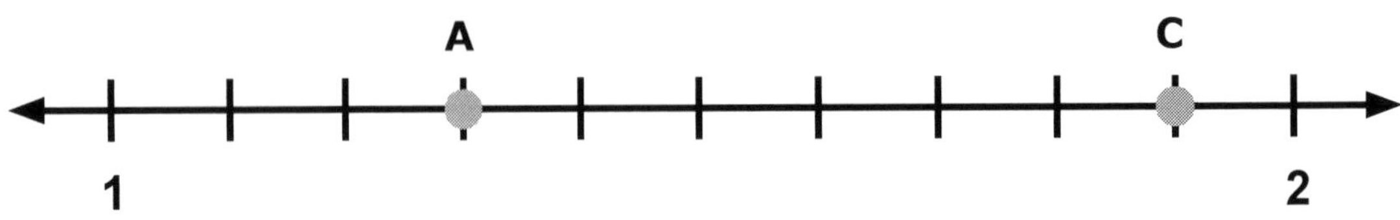

Do you agree with Aditya? Explain your reasoning.

____

____

____

4.NF.C.5

26. Jaylynn has half of a dollar. Marcus has three tenths of a dollar.

Which expression can be used to determine the amount of money they have together?

A. \$0.05 + \$0.03
B. \$0.05 + \$0.30
C. \$0.50 + \$0.30
D. \$0.50 + \$0.03

4.NF.C.6

27. Kai has $\frac{4}{100}$ of a dollar. Silvano has $\frac{8}{10}$ of a dollar.

Which expression can be used to determine the amount of money they have together?

A. $0.04 + $0.08
B. $0.04 + $0.80
C. $0.40 + $0.80
D. $0.40 + $0.08

4.NF.C.6

28. Yesenia, Iris, and Joaquin have a total of $1.00 in coins.

- Yesenia has 3 dimes and 4 nickels
- Iris has 2 dimes, 1 nickel and 3 pennies
- Joaquin has 1 nickel and 17 pennies

Which inequality compares the amount of money Iris has to the amount of money Joaquin has?

A. $0.28< $0.22
B. $0.28 > $0.22
C. $0.50 > $0.28
D. $0.22< $0.50

4.NF.C.7

29. Which statement explains why 0.3 is less than 0.46?

A. The decimal 0.3 has fewer digits than 0.46
B. The number 3 comes before 4 when counting.
C. Three-tenths has 1 less tenth than 0.46.
D. Forty-six is greater than 3.

4.NF.C.7

ASSESSMENT 1

30. True or False: These lines intersect.

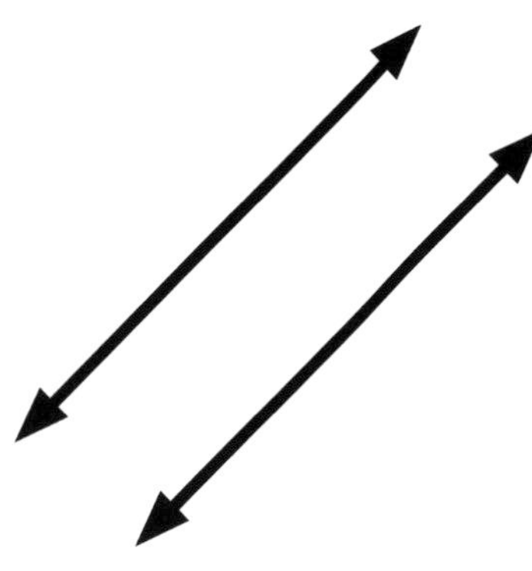

A. True
B. False

(4.G.A.1)

31. True or False: This quadrilateral is a trapezoid.

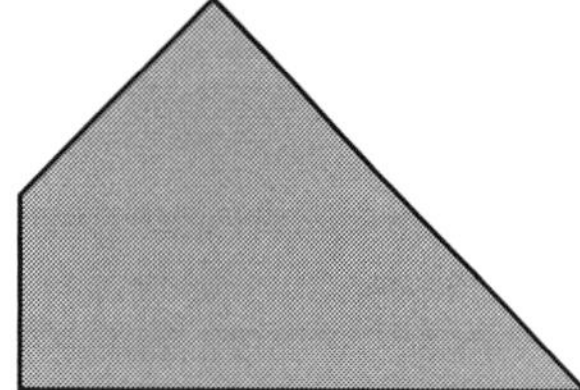

A. True
B. False

(4.G.A.2)

32. True or False: This image has exactly 2 lines of symmetry.

A. True
B. False

(4.G.A.3)

33. Which is more, 20 millimeters or 1 centimeter? ________________

(4.MD.A.1)

34. Which is more, 1,467 grams or 1 kilogram? ________________

(4.MD.A.1)

35. In the following recipe for Guacamole Dip, how many total teaspoons of minced garlic and chopped onion are needed?

- 1 large avocado
- 2 Teaspoons minced garlic
- 1 Tablespoon lemon juice
- 1 Tablespoon chopped onion
- 4 – 6 minced grape tomatoes

A. 6 **B.** 5 **C.** 3 **D.** 2

4.MD.A.2

36. In the following recipe for chili, does the recipe call for more onions or more chili powder?

- 1 lb ground beef
- ¼ cup onions
- 2 ½ cups cooked pintos
- 1 cup water
- 2 cups tomato juice
- ¼ teaspoon garlic powder
- 1 teaspoon salt
- 3 tablespoons chili powder

A. Onions
B. Chili powder
C. An equal amount of both ingredients
D. Not enough information

4.MD.A.2

37. True or False: The perimeter of the following shape is 10 cm.

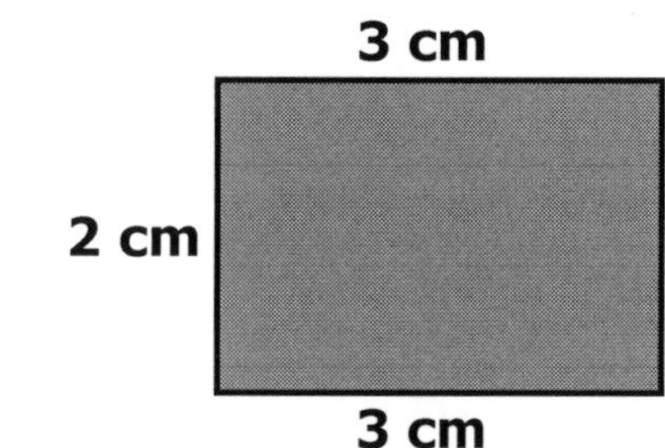

A. True

B. False

4.MD.A.3

38. True or False: Each side of a square playground is 10 meters long. The playground's area is 100 square meters.

A. True **B.** False

4.MD.A.3

39. This line plot shows the growth of the plants in Mrs. Johnson's classroom after one week. How many more plants grew $\frac{1}{2}$ inch than those which grew 1 inch?

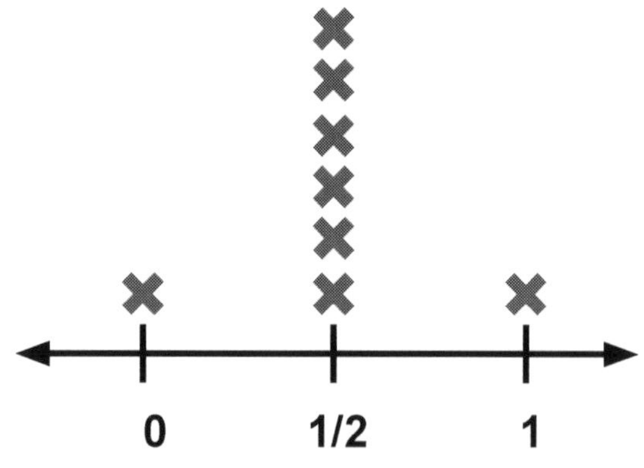

Amount of Growth (inches)

____

4.MD.B.4

40. The arc represents the measure of the angle What is the measure of this angle?

A. 270°
B. 360°
C. 90°
D. 180°

4.MD.C.5.A

41. Robin is turning his plant in a circular motion towards a window to make sure it receives enough sunlight. He turns the plant 10 degrees each hour for 6 hours. Draw a picture of the angle Robin creates.

4.MD.C.5.A

42. Angle ABC is half the measure of Angle ABE. What is the measure of Angle ABC?

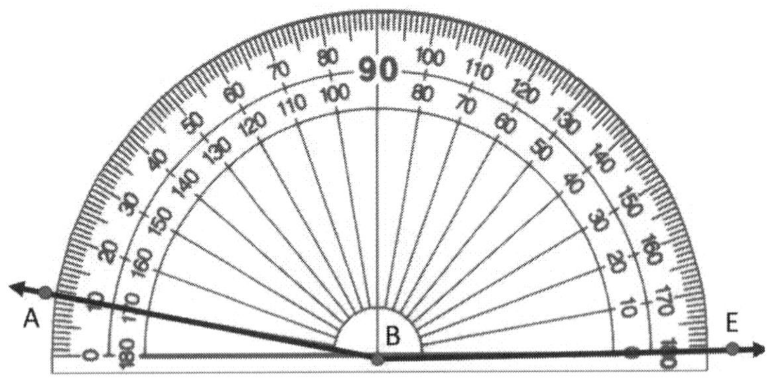

A. 180° **B.** 170° **C.** 110° **D.** 85°

(4.MD.C.6)

43. Yvette draws this ray using a protractor.

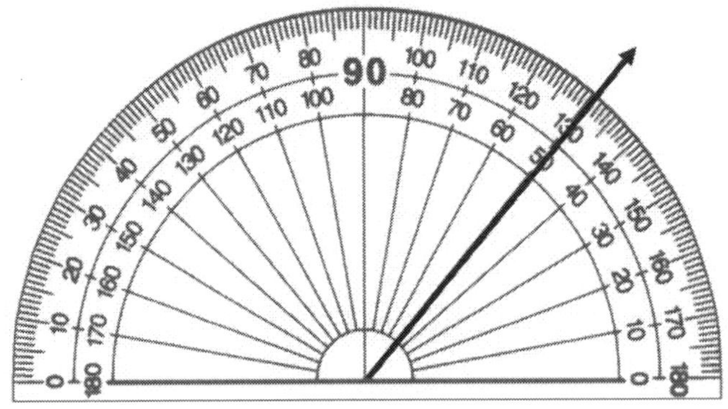

Draw another ray to create a 128° angle.

(4.MD.C.6)

44. What is the measure, in degrees, of Angle h? ____________°

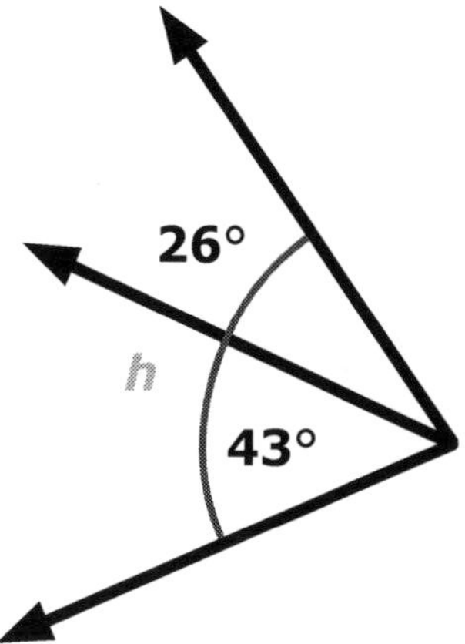

(4.MD.C.7)

45. The measure of Angle LKM is 31 degrees. The measure of Angle JKL is three as large as Angle LKM. What is the measure, in degrees, of Angle JKM?

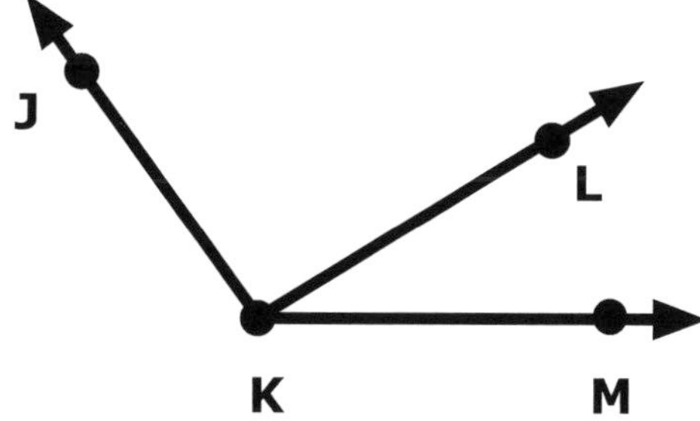

____

(4.MD.C.7)

ASSESSMENT ②

COMPREHENSIVE ASSESSMENTS

1. A fourth-grade class has 28 students. Each student has 3 pencils. How many pencils do the students in the class have?

Write an equation that matches the story. Use p to represent the unknown number.

____

(4.OA.A.1)

2. Rachel raised $75 today at her fundraiser. That is 3 times as much as she raised yesterday. How much money did Rachel raise yesterday?

Rachel raised __________ dollars yesterday.

(4.OA.A.2)

3. There are 6 boxes of colored pencils. Each box has 8 pencils in it. If 9 students each receive the same number of colored pencils, how many pencils will be left over?

____

(4.OA.A.3)

4. Scott has 7 boxes of crayons. Each box has 6 crayons in it. Explain how to express this situation as a multiplication equation.

____

____

____

(4.OA.A.1)

5. Is 3 a factor of 76?

____

(4.OA.B.4)

6. The fourth number in the sequence is 53. The rule is "add 6". What is the first number in the sequence?

____

(4.OA.C.5)

7. There are 5 fourth grade classes at a school. Each class can have up to 29 students. What is the maximum number of fourth graders that could attend the school?

A. 140 **B.** 150 **C.** 34 **D.** 145

(4.OA.A.2)

8. Veronica and her three friends are playing a game called "Guess My Number".

- Veronica's number is 1/100 times Justin's number.
- Justin's number is 1/10 times Eric's number.
- Eric's number is 17,000.
- Brandy's number is 2,000 more than Veronica's number

What are each person's number?

____

____

____

(4.NBT.A.1)

9. Sarah, Monique, and Tara are buying items at the store. Sarah spends 10 times more money than Monique, and Monique spends 1/100 times as much as Tara. Who spends the most money at the store? Explain your reasoning.

____

____

____

(4.NBT.A.1)

10. A fish tank holds 80 fish. Half of the fish are orange and 15 of the fish are blue. The rest of the fish are yellow. How many of the fish are yellow? Explain your thinking.

____

____

(4.OA.A.3)

11. Kami is recording the weight of different whales for a science project.

Whale	Weight (lbs.)
Killer Whale	9,547
Humpback Whale	64,105
Beluga Whale	3,004
Blue Whale	300,019

Which expression represents the expansion of the difference in the weight of the two heaviest whales?

A. 300,000 + 60,000 + 4,000 + 100 + 10 + 4

B. 6,000 + 500 + 40 + 3

C. 300,000 + 60,000 + 4,000 + 100 + 20 + 4

D. 200,000 + 30,000 + 5,000 + 900 + 10 + 4

(4.NBT.A.2)

12. The land area of the United States of America is 9,147,120 square kilometers. How would this number be expressed in written form?

____

(4.NBT.A.2)

13. A 4-digit number has the digits 2, 3, 5 and 7. When rounded to the nearest hundred, the number is labeled as Point A on this number line.

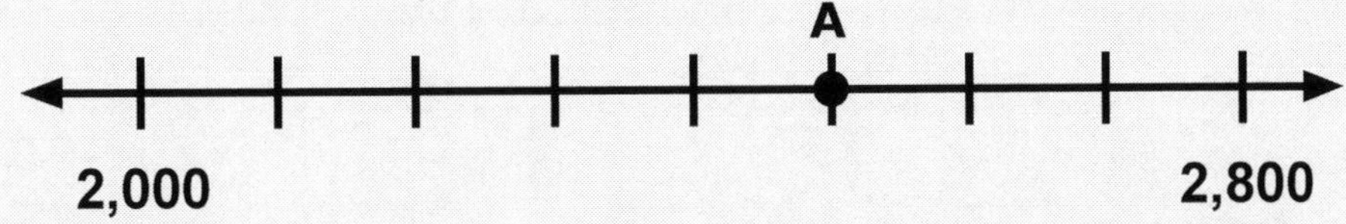

What is the number?

____

(4.NBT.A.3)

14. Zack's teacher asks him to use 5 numbers to write a number that would round to 14,000, when rounded to the nearest hundred.

Zack writes these numbers.

13,954

13,810

14,015

14,495

14,001

Which numbers on Zack's list does not round to 14,000? Explain your reasoning.

____

____

____

(4.NBT.A.3)

15. Write out the steps you would use to subtract these numbers with the standard algorithm.

$$\begin{array}{r} 10{,}030 \\ -\ \ 5{,}159 \\ \hline \end{array}$$

____

____

____

4.NBT.B.4

16. Explain why both strategies can be used to find the product of 11 and 35.

Strategy A: (35×10) + (35×1) = 350 + 35 = 385

Strategy B: (35 ×10) + 35 = 385

____

____

____

4.NBT.B.5

17. Jada has 16 rolls of quarters. Each roll of quarters contains 40 quarters.

How many quarters does Jada have? ____________

4.NBT.B.5

18. Lyndon uses this strategy to divide two numbers.

$$450 \div 8$$
$$(480 \div 8) = 60$$
$$480 - 450 = 30$$
$$60 - 30 = 30$$

Do you agree with Lyndon? Explain your reasoning.

____

____

____

4.NBT.B.6

19. Bethany uses this model to represent a multiplication situation.

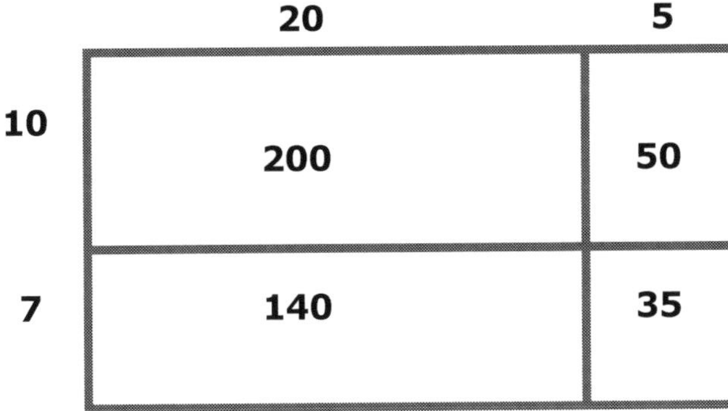

What is the value of the product?

A. 25 **B.** 17 **C.** 425 **D.** 200

4.NBT.B.6

20. Ralph is counting his sports cards. Out of 20 cards, 5 are baseball cards, and the rest are soccer cards. Which fraction is equivalent to the fraction of soccer cards in Ralph's card collection?

A. $\frac{3}{4}$ **B.** $\frac{2}{5}$ **C.** $\frac{4}{10}$ **D.** $\frac{5}{15}$

(4.NF.A.1)

21. Which list shows fractions listed from greatest to least?

A. $\frac{4}{7}$ $\frac{5}{8}$ $\frac{1}{6}$ **B.** $\frac{6}{7}$ $\frac{3}{8}$ $\frac{1}{4}$ **C.** $\frac{3}{5}$ $\frac{1}{3}$ $\frac{8}{10}$ **D.** $\frac{6}{9}$ $\frac{2}{6}$ $\frac{1}{2}$

(4.NF.A.2)

22. Amy shared a pie with her friends. Erin ate $\frac{1}{6}$ of the pie, Harry ate $\frac{2}{6}$ of the pie, and Amy ate $\frac{1}{6}$ of the pie. Write an expression, with 4 identical fractions, to represent the total amount of pie they ate.

____

(4.NF.B.3)

23. Wanda walks $\frac{3}{4}$ miles each day for 8 days. She uses this equation to represent the total number of miles she walks.

$$24 \times \frac{1}{4} = \frac{24}{4} = 6$$

Do you agree with Wanda? Explain your reasoning.

____

____

____

(4.NF.B.4)

24. Miriam places cubes side by side to measure the length of a line. Each cube is $\frac{3}{100}$ centimeters wide. Miriam determines the line is $\frac{30}{100}$ centimeters long and writes this expression to show her thinking.

$$\frac{3}{100}+\frac{3}{100}+\frac{3}{100}+\frac{3}{100}+\frac{3}{100}+\frac{3}{100}+\frac{3}{100}+\frac{3}{100}+\frac{3}{100}+\frac{3}{100}=\frac{3}{100}$$

How many cubes does Miriam use to measure the length of the line? Explain your reasoning.

____

____

____

(4.NF.C.5)

25. A bag contains red, blue, green, and black marbles. This table shows the amount of marbles represented as a fraction.

Marble Color	Red	Blue	Green	Black
Fraction of Bag	$\frac{5}{10}$	$\frac{30}{100}$	$\frac{10}{100}$	$\frac{1}{10}$

Zoe says most of the bag is blue and green marbles. Do you agree with Zoe? Explain your reasoning.

____

____

____

(4.NF.C.5)

26. Itzel has $\frac{34}{100}$ of a dollar. Henrietta has $\frac{4}{10}$ of a dollar.

How much more money does Henrietta have than Itzel?

A. \$0.30 **B.** \$0.38 **C.** \$0.06 **D.** \$0.60

(4.NF.C.6)

27. Kiva's cat can jump $5\frac{7}{10}$ feet. How would this distance be expressed as a decimal?

A. 5.07 **B.** 5.7 **C.** 5.710 **D.** 5.3

(4.NF.C.6)

28. What decimal represents the number of squares, as a portion of the whole figure, needed for the value of this model to be equivalent to 0.5? Explain your reasoning.

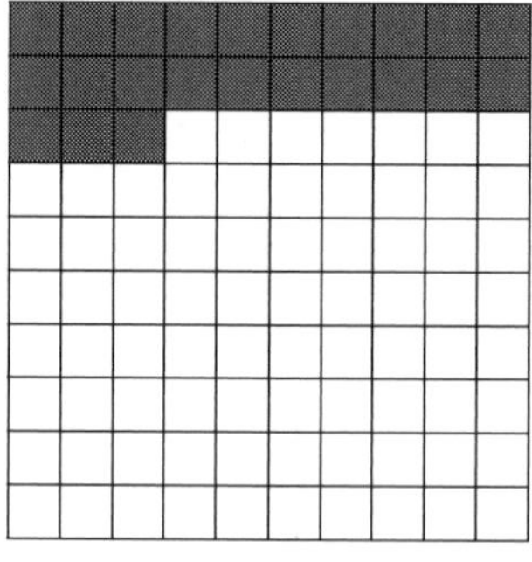

____

____

____

(4.NF.C.7)

29. Nelson cuts this pie into fifths. He eats $\frac{3}{5}$ of the pie.

Then, he cuts this pie into eighths. He gives his friends $\frac{4}{8}$ of the pie.

Nelson knows $\frac{3}{5}$ is equivalent to 0.60, and $\frac{4}{8}$ is equivalent to 0.50. He decides he has eaten more pie than his friends ate.

Do you agree? Explain your reasoning.

____

____

____

4.NF.C.7

30. True or False: These lines are intersecting.

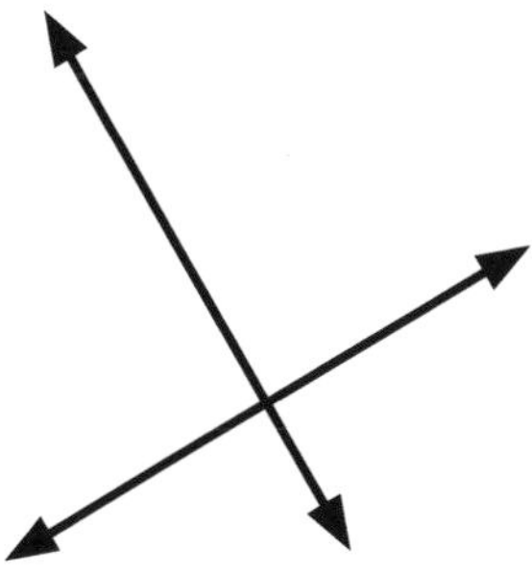

A. True
B. False

4.G.A.1

31. What kind of triangle, classified by its angles, is this triangle?

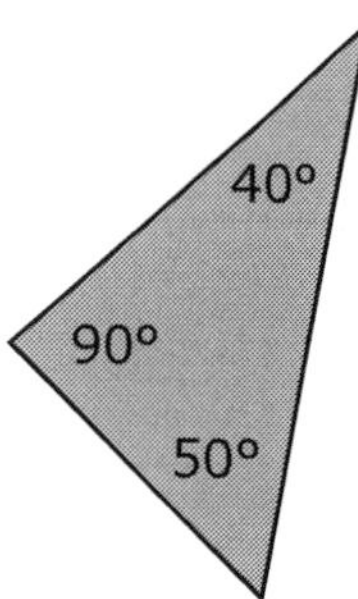

____

(4.G.A.2)

32. How many lines of symmetry are in this shape?

____

(4.G.A.3)

33. True or False: 1 liter = 3,000 milliliters

A. True **B.** False

(4.MD.A.1)

34. True or False: The length 3,998 millimeters is greater than the length 4 meters.

A. True **B.** False

(4.MD.A.1)

35. The coffee shop is 2.6 kilometers west of the bakery and 8.7 kilometers east of the game store. How many meters apart are the game store and the bakery?

____

(4.MD.A.2)

36. The toy store is 5.9 kilometers west of the library, and the library is 3,250 meters west of the train station. How far is the toy store from the train station, in meters?

____

(4.MD.A.2)

37. The rectangles below have the same perimeter. If the area of the horizontal rectangle is 28 square meters, what are its dimensions, in meters?

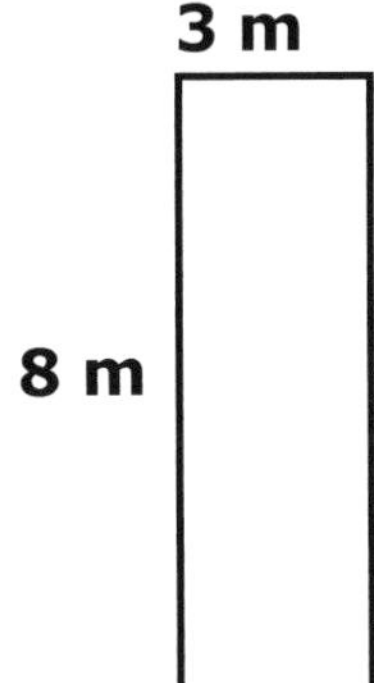

____

____

____

(4.MD.A.3)

38. A square flower garden has sides that are 9 meters long. Fencing costs $5.00 per meter. How much would it cost to put a fence around the flower garden? $ __________

(4.MD.A.3)

39. This line plot shows the amount of wax inside 8 candle jars. How many candle jars have more than $\frac{1}{2}$ of a cup of wax?

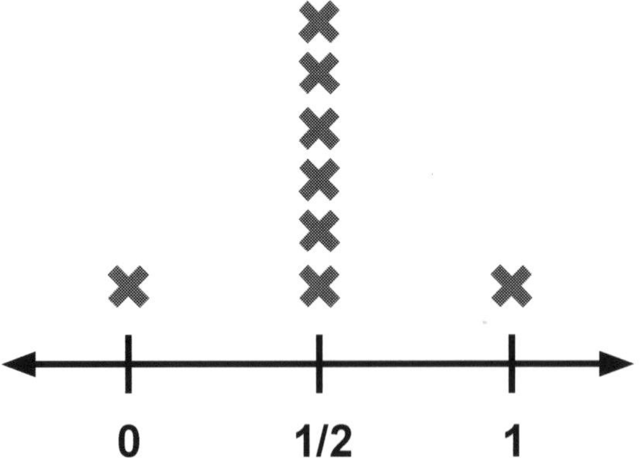

____

____

____

(4.MD.B.4)

40. Which fraction represents the closest estimate of the portion of a full circle represented by this angle?

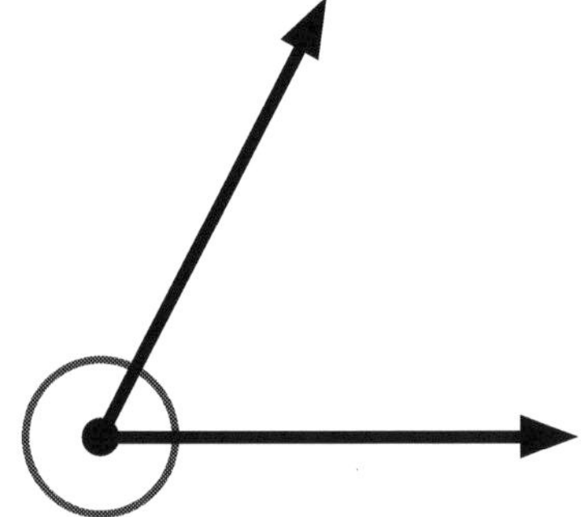

A. $\frac{70}{360}$ **B.** $\frac{45}{360}$

C. $\frac{90}{360}$ **D.** $\frac{120}{360}$

(4.MD.C.5.A)

41. This timer turns 1-degree counterclockwise each second.

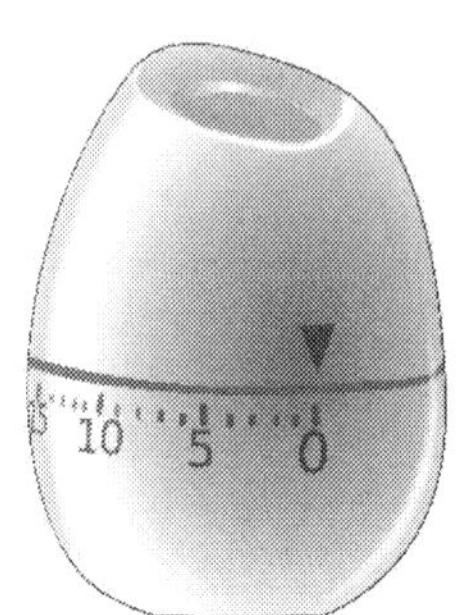

Eric turns the arrow on the timer to the 35 second mark. How many degrees does the arrow move?

____

(4.MD.C.5.B)

42. Angle ABC is 3 times as large as angle ABF. What is the measure of Angle ABC, in degrees?

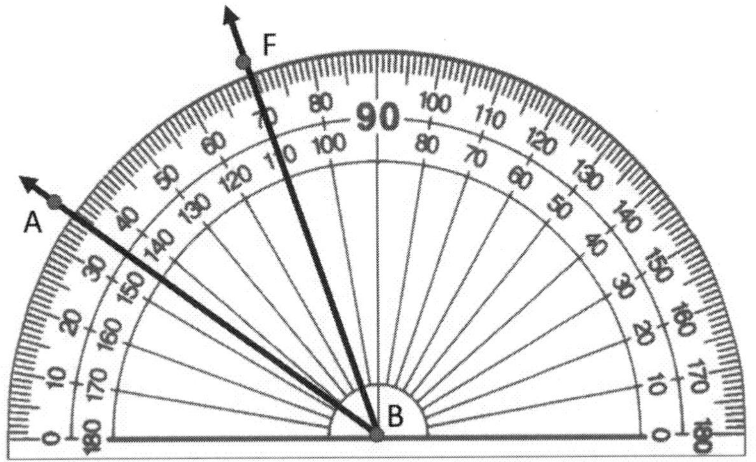

(4.MD.C.6)

43. Mannie uses this protractor to measure an angle.

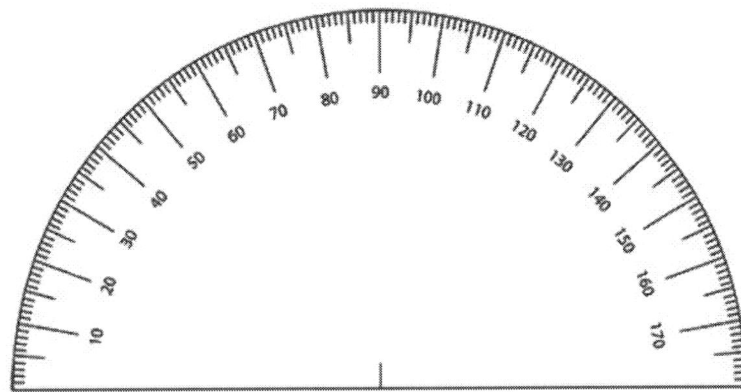

One ray of the angle crosses from the center through the 22°. The second ray passes from the center through the 131°. What is the measure of the angle?

A. 153° **B.** 109°
C. 131° **D.** 111°

(4.MD.C.6)

ASSESSMENT 2

44. What is the measure of Angle ***d***?

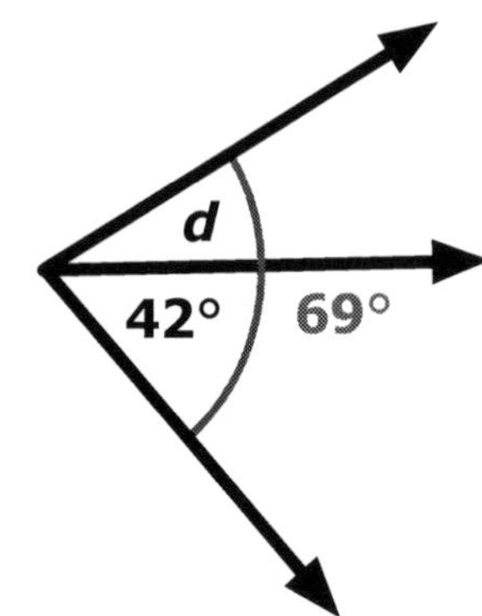

A. 42°
B. 27°
C. 30°
D. 111°

(4.MD.C.7)

45. This diagram shows the angle at which the sun appears in the sky at different times throughout the day.

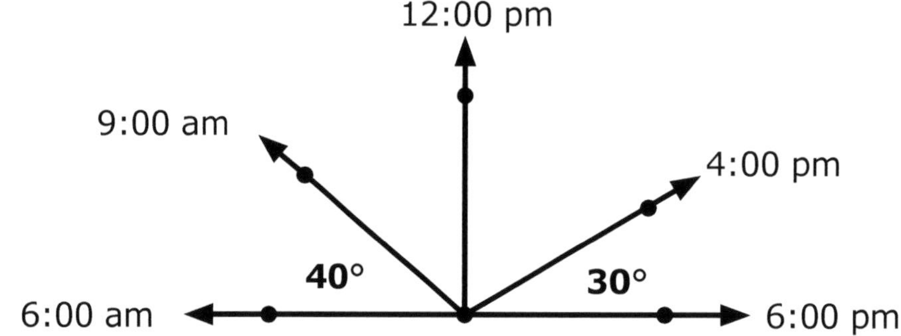

How many degrees does the sun move between 9am and 4pm?

____

(4.MD.C.7)

ANSWERS AND EXPLANATIONS

Operations & Algebraic Thinking
Problem Solving All Operations 206
Factors and Multiples 207
Patterns 208
Chapter Review 209
Extra Practice 210

Number & Operations in Base Ten
Place value - rounding and Comparing 211
Add and subtract 212
Multiply and Divide 214
Chapter Review 215
Extra Practice 216

Number & Operations - Fractions
Equivalent Fractions and Ordering Fractions 218
Fraction Operations 219
Decimal Conversion and Comparison 221
Chapter Review 223
Extra Practice 224

Measurement & Data
Problem Solving - Conversion of Measurements 226
Graphs and Data Interpretation 227
Geometry - Concepts of Angles 230
Chapter Review 231
Extra Practice 232

Geometry
Lines and Angles 233
Two-Dimensional Shapes 234
Line of Symmetry 235
Chapter Review 236
Extra Practice 237

Assessment 1 239

Assessment 2 241

OPERATIONS & ALGEBRAIC THINKING
UNIT 1: PROBLEM SOLVING ALL OPERATIONS

1 Answer: A

Explanation: The equation that multiplies 15 and 6 shows that the orchard has 15 bags of small apples and 6 times that number of large apples.

2 Answer: A

Explanation: Julie has twice as much as 14. Twice means 14 times 2.

3 Answer: D

Explanation: The equation tells that 72 is 6 times as many as 12 and 12 times as many as 6.

4 Answer: D

Explanation: The number of cookies sold equals 7 times 9.

5 Answer: C

Explanation: The equation shows the cost is 3 times 4 or 4 times 3.

6 Answer: I agree. The numbers can be multiplied in any order.

Explanation: Answers must include a reasonable explanation such as, "I agree. You can multiply numbers in any order and get the same answer. Multiplying 6 by 8 or multiplying 8 times 6 both equal 48."

7 Answer: 45 = 5 x 9 or 45 = 9 x 5

Explanation: Answers must include a reasonable explanation such as, "Joseph should write 45 = 5x9. Five times as many as 9 is the same as 5 groups of 9 so I need to multiply."

8 Answer: 78

Explanation: Answers must include a reasonable explanation such as, "Jacob is thinking of the number 78 because 13 groups of 6 equals 78."

9 Answer: 13

Explanation: Answers must include a reasonable explanation such as, "John picked 13 yellow flowers. I solved by dividing 65 by 5, which is 13."

10 Answer: 15 times 5 is 75.

Explanation: Answers must include a reasonable explanation such as, "I would explain to him that 15 times 5 is 5 groups of 15. He could write 15 five times and add them together."

11 Answer: C

Explanation: The total number of kindergartners is 21 times 4.

12 Answer: D

Explanation: To find how many points were scored in the second quarter, multiply 21 by 4.

13 Answer: 240

Explanation: 60x4 = 240

14 Answer: 42

Explanation: 7x6 = 42

15 Answer: A

Explanation: Multiply 5 times 17, which equals 85. Then, subtract 80 from 85, which is 5. She has $5 left.

16 Answer: B

Explanation: Add 53 + 41 = 94.Then, divide 94 by 10, which is 9 with a remainder of 4.

17 Answer: B

Explanation: Each sheet contains 11 problems. Multiply 30x11 which is 330. Estimate the total by rounding 330 down to 300, meaning that the total number is a little more than 300.

18 Answer: B

Explanation: Add 60 + 60. He counted 120 spots. Divide 120 by 6. He counted the spots on 20 ladybugs.

19 Answer: $3

Explanation: Each child received $15 because 60 divided by 4 is 15. Since the toy costs $18 and she has $15, she needs 3 more dollars.

20 Answer: 20

Explanation: Multiply 31 x 2 = 62. Then, 82 − 62 = 20.

OPERATIONS & ALGEBRAIC THINKING
UNIT 2: FACTORS AND MULTIPLES

1 Answer: A

Explanation: The number 61 is a prime number because it cannot be divided by any number except 1 and itself.

2 Answer: C

Explanation: The number 71 is the next prime number after 67 because it cannot be divided by any number except 1 and itself.

3 Answer: D

Explanation: The number 21 is not prime because it can be divided by 3 and 7.

4 Answer: C

Explanation: The number 22 has 2 factor pairs: 1 & 22 and 2 & 11.

5 Answer: D

Explanation: The number 44 cannot be divided equally by 6. Dividing 44 by 6 leaves a remainder of 2.

6 Answer: B

Explanation: The factors of 98 are 2, 7, 14, and 49.

7 Answer: D

Explanation: The factor pairs of 88 are 1 & 88, 2 & 44, 4 & 22, and 8 & 11.

8 Answer: D

Explanation: The number 72 divided by 4 results in 18, with no remainder. Thus, 4 is a factor.

9 Answer: B

Explanation: The number 86 divided by 3 results in 28 and a remainder of 2. Thus, 3 is not a factor of 86.

10 Answer: C

Explanation: Dividing 96 by 6 results in 16 with no remainder. Therefore, 6 is a factor of 96.

11 Answer: Prime

Explanation: Answers must include a reasonable explanation such as, "The number 37 is a prime number because its only factors are 1 and 37."

12 Answer: 19

Explanation: After 17 is 18, then 19. All even numbers greater than 2 are composite numbers. The number 18 can be divided by 2, and 19 can only be divided evenly by 1 and 19 so it is prime."

13 Answer: 1 & 62—2 & 31

Explanation: Answers must include a reasonable explanation such as, "The factor pairs for 62 are 1 & 62 and 2 & 31 because 31 is a prime number so no other factor pairs of 62 exist.

14 Answer: No

Explanation: Divide 94 by 3. The result is 31 with a remainder of 1. Therefore, 3 is not a factor of 94.

15 Answer: Composite

Explanation: All composite numbers have more than one factor pair. Prime numbers have only one factor pair.

16 Answer: 4

Explanation: The prime numbers between 1 and 10 are 2, 3, 5, and 7.

17 Answer: 68

Explanation: The number 67 is prime so 68 is the next composite number after 66.

18 Answer: Prime

Explanation: The number 89 is prime because its only factor pair is 1 & 89.

19 Answer: Prime

Explanation: Prime numbers have exactly 2 factors, 1 and the number itself.

20 Answer: 3

Explanation: The prime numbers between 70 and 80 are 71, 73 and 79.

OPERATIONS & ALGEBRAIC THINKING UNIT 3: PATTERNS

1 Answer: C

Explanation: Multiply the number of hours by 3. For 100 hours, the number of trees trimmed is 300 because $100 \times 3 = 300$.

2 Answer: B

Explanation: The shape turns half way around each time.

3 Answer: D

Explanation: The rule of the pattern is multiply the last number by 2 to find the next number.

4 Answer: C

Explanation: The ones digit will be 0, 5, 0, 5, and so on. This means that the numbers will alternate between even and odd.

5 Answer: A

Explanation: The first number is even and the number being added is even, so all of the numbers will be even.

6 Answer: B

Explanation: The rule of the pattern is add 9. The next number is 36.

7 Answer: C

Explanation: The ones digit in the first number is odd and the number being added is an odd number. That means the numbers will be odd, then even, then odd, then even, and so on.

8 Answer: 20

Explanation: The rule is add 4. The missing number is 20 because $16 + 4 = 20$..

9 Answer: 76

Explanation: The rule is add 5. The missing number is 76 because $71 + 5 = 76$.

10 Answer: Add 12

Explanation: Using the pattern, $60 - 48 = 12$, $72 - 60 = 12$, so the rule of the sequence is add 12.

11 Answer: 94

Explanation: Starting with 64 and adding 10, the sequence is 64, 74, 84, 94, ... The fourth number is 94.

12 Answer: 25

Explanation: Since the first term is 1 and the second term is 5, the rule is to multiply by 5. Confirming with the last two terms: $125 \times 5 = 625$.

13 Answer: 60

Explanation: The rule is add 15 (or subtract 15 if moving from right to left) so the number before 75 is 60.

14 Answer: 190

Explanation: The sequence starts with 55, so $55 \times 2 = 110$, and $110 - 10 = 100$. Then, $100 \times 2 = 200$, and $200 - 10 = 190$.

15 Answer: 80

Explanation: The rule is to divide by 2 because $40 / 2 = 10$. Therefore, half of 160 is 80.

16 Answer: 61

Explanation: The first number in the sequence is 85; $85 - 12 = 73$. The second number is 73; $73 - 12 = 61$. The third number is 61.

17 Answer: 56

Explanation: If the rule of the sequence is to add 11, then, the rule going to the left of the given number is to subtract 11. Beginning with the 3rd term of 78; $78 - 11 = 67$, and $67 - 11 = 56$. The first term of the sequence is 56, the second term is 67, and the third term is 78 as given.

18 Answer: Even.

Explanation: Adding an even number to other even numbers results in even numbers.

19 Answer: Add 4

Explanation: The pattern is add 4 because each number increases by 4.

20 Answer:

Explanation: The circle is turning a quarter of the way around clockwise. The next time the circle turns, the open part will be on the lower left side of the circle.

OPERATIONS & ALGEBRAIC THINKING CHAPTER REVIEW

1 Answer: B

Explanation: The equation shows a number x times 9 equals 36. The variable x is Kate's age.

2 Answer: A

Explanation: The equation $4 \times 3 = s$ shows that the number s is 3 times 4.

3 Answer: D

Explanation: The equation $84 = 14 \times 6$ means that 84 is 14 times 6.

4 Answer: C

Explanation: The comparison 63 is 7 times 9 is represented by the equation $63 = 9 \times 7$.

5 Answer: B

Explanation: The number 50 has 3 factor pairs (1 & 50, 2 & 25, 5 & 10).

6 Answer: A

Explanation: Multiplying 2 by 29 equals 58.

7 Answer: 1

Explanation: Prime numbers can only be factored by 1 and the number itself. Thus, 1 is the only other factor of a prime number.

8 Answer: Composite

Explanation: The factors of 91 are 1, 7, 13, and 91. Thus, 91 is not prime.

9 Answer: C

Explanation: The relationship is 15 times 3 equals 45.

10 Answer: B

Explanation: Multiply 15 times 9 resulting in 135.

11 Answer: B

Explanation: Multiply 8 times 3 and get 24.

12 Answer: 135

Explanation: Find the maximum books by multiplying 27 times 5 which equals 135.

13 Answer: B

Explanation: The pattern is "subtract 35" so the next number is $280 - 35 = 245$.

14 Answer: B

Explanation: The pattern is $22 + 11 = 33$, $33 + 11 = 44$. The rule of the pattern is "add 11".

15 Answer: B

Explanation: From left to right, the rule is to multiply by 2. Working backwards (right to left), the rule become divide by 2. The number 64 divided by 2 is 32.

16 Answer: 3

Explanation: The rule is divide by 3. The next number in the sequence is 3.

17 Answer: B

Explanation: Calculate 45 minus 10 which is 35. Then, 35 divided by 7 is 5.

18 Answer: 39

Explanation: Order of operations says divide 75 by 5 first giving 15. Then, add 24 to 15: $24 + 15 = 39$.

19 Answer: 13

Explanation: Add 42 plus 38 and gets 80. Then, divide 80 by 6 which equals 13 with a remainder of 2. He can make 13 full packs of 6 fish in each pack.

20 Answer: 70

Explanation: The packs contain 5 times 12 or 60. Then, 130 minus 60 is 70. The bakery will not package 70 cupcakes.

OPERATIONS & ALGEBRAIC THINKING EXTRA PRACTICE

1 Answer: 90 = 9 x 10
or
90 = 10 x 9

Explanation: The equation shows 90 as 10 times 9 or 9 times 10.

2 Answer: 8 x 11 =88
or
11 x 8 = 88

Explanation: The equation shows that 8 groups of 11 equals 88 or 11 groups of 8 equals 88.

3 Answer: 24
s = 4 x 6

Explanation: There are 24 students playing in the playground. The equation that shows that the number of students s on the playground is equals 6 times 4.

4 Answer: 90 = 6 x 15
or
90 = 15 x 6

Explanation: The equation must contain 15 times 6. The equation shows that 90 is 15 times 6.

5 Answer: 5

Explanation: Multiplication can be performed in any order. Thus, 4 times 5 gives the same answer as 5 times 4.

6 Answer: J = 7 x 4

Explanation: The equation must use the variable J and shows that J is equal to 7 times 4.

7 Answer: 60 = 12 x 5
or
60 = 5 x 12

Explanation: The number 60 is equal to 12 times 5. The multiplication can also be 5 times 12.

8 Answer: Agree

Explanation: The number 99 divided by 11 is 9 with no remainder.

9 Answer: 24

Explanation: Julie has 24 shirts because 4 groups of 6 is 24.

10 Answer: 96

Explanation: There is room for 96 bikes on the bike rack because 8 groups of 12 is 96.

11 Answer: 81
81 = 9 x 9

Explanation: Ms. Thompson needs 81 juice boxes because 9 times 9 is 81. The equation is 81 = 9 x 9.

12 Answer: 5

Explanation: Find the answer using 18 times 5 equals 90.

13 Answer: Add 5
or the numbers are multiples of 5

Explanation: The pattern is increasing by 5.

14 Answer: A

Explanation: The rule of the pattern is multiply by 3. Each term is 3 times the previous term.

15 Answer: 21

Explanation: The first number is 5. Adding 4 gives 9 as the second term. Adding 4 gives 13 as the third term. Adding 4 gives 17 as the fourth term. Adding 4 gives 21 as the fifth term. Answers must include a reasonable explanation such as, "The pattern would be 5, 9, 13, 17 then 21. The 5th number in the pattern is 21."

16 Answer: Multiply by 10

Explanation: The pattern is to multiply

each number by 10. Each term is 10 times the term before it.

17 Answer: 21

Explanation: The number is 21. 56 divided by 8 is 7, and 7 times 3 is 21.

18 Answer: Yes. The product of 49 x 5 can be estimated by multiplying 250 x 5.

Explanation: Araceli's estimate is reasonable because 49 is very close to 50. Then, 50 times 5 is 250. Thus, 49 times 5 will just be a little less than 50 times 5.

19 Answer: A

Explanation: Add 35 and 62 to get 97. Then, 97 divided by 7 is 13 with some left over so he can make 13 pies.

20 Answer: A

Explanation: Start with 4 times 30 and get 120. Then, 120 minus 72 is 48.

NUMBER & OPERATIONS IN BASE TEN UNIT 1: PLACE VALUE - ROUNDING AND COMPARING

1 Answer: A

Explanation: The digit 4 in 243,188 represents 40,000. The digit 4 in 947 represents 40. Then, 40,000/40 = 1,000. Thus, the digit 4 in 243,188 is 1,000 times the digit 4 in 947.

2 Answer: D

Explanation: The 6 in 63,143 is in the 10,000s digit. The 6 in 4,006 is in the ones digit. Therefore, the digit 6 in 63,143 is 10,000 times the 6 in 4,006.

3 Answer: C

Explanation: The ratio of students to classes is 180:20. The numbers in this ratio can both be divided by 2, 4, 5, 10, 0r 20. Choice C shows the result of dividing both numbers by 10. Since each class has 1 teacher, the ratio of students to classes is the same as the ratio of students to teachers.

4 Answer: D

Explanation: The ratio of 350 pencils : groups of 50 pencils is 350:50. Both numbers can be divided by 2, 5, 10, 25, or 50. The ratio Choice D shows the result of dividing both numbers by 10.

5 Answer: 20

Explanation: Use the expression 2800/140. Divide 2800 by 140 and get 20.

6 Answer: 4,200

Explanation: The book has 42 x 100 words, which is 4,200.

7 Answer: 370

Explanation: Ten times 37 is 370.

8 Answer: D

Explanation: Expanded form is writing a number to show the value of each digit. The number is shown as a sum of each digit multiplied by its matching place value. Calculate each number in the expanded form by multiplying the digit by its place value.

9 Answer: B

Explanation: Expanded form is writing a number to show the value of each digit. The number is shown as a sum of each digit multiplied by its matching place value. Calculate each number in the expanded form by multiplying the digit by its place value.

10 Answer: A

Explanation: The numbers given in the stem are 1,405 and 780. Find the total distance by adding them, giving 2,185 miles.

11 Answer: 30,811 < 30,901

Explanation: Add 3,800 to the cost of each car which results in 30,901 and 30,811. The inequality has the smaller number on the left.

12 Answer: $720

Explanation: Wes has $2,806 and Mary has $2,086. If they split the cost of the television, each of them will pay $245. Wes

will have $2,561 and Mary will have $1,841 left. The difference in the amount of money they will have left is $720.

13 Answer: No

Explanation: The value of the digit 8 should be 8,000 not 8,000,000.

14 Answer: No

Explanation: The value of the digit 4 should be 40 not 400.

15 Answer: A

Explanation: The number 425, rounded to the nearest hundred is 400 because the tens digit is 2. The 4 is in the 100s digit. When the tens digit is less than 5, all lower digits become 0.

16 Answer: B

Explanation: To estimate by rounding, round the values in the problem before doing the math operation. The number 957, rounded to the nearest hundred, is 1,000. The number 648, rounded to the nearest hundred, is 600.

17 Answer: 140 + 30 + 40 = 210

Explanation: To estimate by rounding, round the values in the problem before doing the math operation. After receiving the additional items, Bebe has 130 + 10 toys, 30 + 0 video game, and 30 + 10 stuffed animals. After rounding, she has 140 toys, 30 video games, and 40 stuffed animals. The equation is 140 + 30 + 40 = 210.

18 Answer: 200

Explanation: Jacki's candy, rounded to the nearest hundred, is 100 pieces. Jordan's candy, rounded to the nearest hundred, is also 100. Together, they have approximately 200 pieces of candy.

19 Answer: 67,400 rounds to 70,000.

Explanation: Rounding to the nearest 10,000 changes 67,400 to 70,000 because the 1,000s digit is 7, which rounds up.

20 Answer: Irving

Explanation: Rounding 18,999 to the nearest ten-thousand would result in 20,000 because the 100s digit is 9, and rounds up, so add 1 to 18.

NUMBER & OPERATIONS IN BASE TEN
UNIT 2: ADD AND SUBTRACT

1 Answer: B

Explanation: The standard algorithm for addition involves adding numbers according to their value (base-10). In this situation, no regrouping is necessary.

2 Answer: C

Explanation: The standard algorithm for addition involves adding numbers according to their value (base-10). In this situation, the ones and tens places needed to be regrouped.

3 Answer: D

Explanation: The standard algorithm for addition involves adding numbers according to their value (base-10). In this situation, the values in the ones and hundreds places needed to be regrouped before subtraction could take place.

4 Answer: C

Explanation: Using the standard algorithm and starting with the values in the ones place would require regrouping. Instead of subtracting 8 from 2, regrouping allows subtraction of 8 from 12.

5 Answer: A

Explanation: Using the standard algorithm and starting with the values in the ones place would require regrouping. Instead of subtracting 6 from 0, regrouping allows subtraction of 6 from 10.

6 Answer: C

Explanation: The standard algorithm for addition involves adding numbers according to their value (base-10). In this situation, the ones, tens, and hundreds will need to be regrouped during the addition because their sums are 10 or more.

7 Answer: B

Explanation: Using the standard algorithm and starting with the values in the ones place would require regrouping. As a first step, instead of subtracting 7 from 5, regrouping allows subtraction of 7 from 15. Then, regroup in the 10s place and in the 100s place.

8 Answer: A

Explanation: Expression A requires regrouping of the ones and the tens because the sums of both digits are 10 or more.

9 Answer: Yes

Explanation: The number of tens, when added, will be 10 or greater, which means they should be regrouped as hundreds.

10 Answer: 3

Explanation: The digits in the ones, tens, and hundreds place in the minuend are less than the digits in the subtrahend. Therefore, each digit must be regrouped before subtracting.

11 Answer: 5,347 – 3,659 = ______

Explanation: Since the values in the ones place had to be regrouped, this means the minuend has a 7 in the ones place and the subtrahend has a 9 in the ones place. Since the hundreds and thousands places also had to be regrouped, and 5 tens are being subtracted from 13 tens, this means the minuend has a 4 in the tens place and the subtrahend has a 5 in the tens place. This same idea can be applied to the hundred and thousands place.

12 Answer: Answers may vary

Explanation: The student should describe the process for regrouping. To subtract 374 – 176, regrouping is needed to subtract the ones and the tens.

13 Answer: B

Explanation: Expression B does not require regrouping of the ones but does not require regrouping of the tens, the hundreds, and the thousands.

14 Answer: A

Explanation: Equation A does not require regrouping of the ones, but dues require regrouping of the hundreds into tens and regrouping the thousands into hundreds.

15 Answer: C

Explanation: Equation C requires regrouping of all digits in order to subtract.

16 Answer: Yes. Explanations may vary.

Explanation: The number of ones, tens, and hundreds in the minuend are all zeros and are less than the number of ones, tens and hundreds in the subtrahend. Therefore, regrouping is necessary to subtract the first three digits.

17 Answer: Yes

Explanation: Yes. She regrouped for each place value position where the sum is greater than 9.

18 Answer: Answers may vary

Explanation: The student should describe the process for regrouping, which is needed for the ones, the tens, and the hundreds digits.

19 Answer: B

Explanation: When adding the two numbers, only the sum of the tens digit gives a value greater than 9 and requires regrouping into a different place.

20 Answer: 1,398 + 2,525 = ______

Explanation: Based on the numbers in the problem and the regroupings, these two numbers are being added:
1,398 + 2,525 = ______

NUMBER & OPERATIONS IN BASE TEN
UNIT 3: MULTIPLY AND DIVIDE

1 Answer: B

Explanation: Solve this problem with multiplication: 45 times 12 is 540. (45x12=540)

2 Answer: D

Explanation: Solve this problem with multiplication: 27 times 18 is 486. (27x18=486)

3 Answer: C

Explanation: Using a rounding and decomposition strategy, 28x11 can be calculated as (30x11)-(2x11).

4 Answer: A

Explanation: The lowest possible number of points for the season is 15x18 or 270 points. The highest possible number of points for the season could be 15x35 or 525 points. 345 points is between these 2 values.

5 Answer: B

Explanation: The lowest possible amount is 14x12=168 ounces. The highest possible amount is 14x18=252 ounces. Roseanne's birds eat between 168 ounces and 252 ounces of food during a two week period.

6 Answer: $2,150

Explanation: Multiplying half the student population (1,075) by the cost of each lunch ($2) is $2,150.

7 Answer: 3,822

Explanation: A variety of strategies may be used to find the value of the expression. Multiply 78 by 49.

8 Answer: $1,140

Explanation: One-third of 90 is 30. There are 30 adults and 60 children in the group. Using the expression (30x18)+ (60x10), the total amount spent is $1,140.

9 Answer: Yes

Explanation: Jamal rewrites the expression using partial products, and then correctly adds all the products together.

10 Answer: Answers may vary

Explanation: The area of the flower bed is 44 square feet. If she plants 12 seeds in each square foot, she plants 44x12 seeds or 528 seeds.

11 Answer: B

Explanation: Divide 249 by 12 which gives 20 with a remainder of 9. The farmer fills 20 baskets with 20 pears each, and one basket with 9 pears. So, he needs 21 baskets.

12 Answer: C

Explanation: Divide 271 by 21 which gives 12 with a remainder of 19. Steven fills 12 shelves with 21 books on each of them. Then he fills another shelf with 19 books. He needs 13 shelves.

13 Answer: B

Explanation: There are enough pencils, erasers, and pens to create 43 complete bags. The limiting item is the pencils.

14 Answer: D

Explanation: The solution is determined by decomposing 540 into 2 values which are divisible by 6: 540 can be decomposed into 540+6. Then, divide each value by 6.

15 Answer: B

Explanation: The solution is determined by decomposing 378 into 2 values which are divisible by 9: 378 can be decomposed into 360+18. Then, divide each value by 9.

16 Answer: B

Explanation: The number 2,104 can be decomposed into 2,100+4. 2,100 can be evenly divided by 7; the quotient is 300. The remainder is 4.

17 Answer: 24

Explanation: The number 219 divided by 9 is 24, with a remainder of 3. This means there will be 24 complete teams.

18 Answer: 223

Explanation: Divide 892 by 4. The quotient is 223.

19 Answer: 375

Explanation: The team sells 1,482 towels for $6 each. Multiplying these two values gives $8,916. Subtracting $8,916 from the total amount earned ($12,267) gives the amount of money earned from selling baseball bats ($3,375). Divide $3,375 by 9 to find the number of bats sold. The team sold 375 bats.

20 Answer: 1,290 ÷ 6

Explanation: Divide the amount of money collected by the price of each ticket. The school sold 215 tickets.

NUMBER & OPERATIONS IN BASE TEN CHAPTER REVIEW

1 Answer: B

Explanation: The given expression can be written as 720/8 by dividing both numbers by 100. Then, divide 720 by 8 which gives 90. The quotient of 72 and 8 is 9, so the given expression is 10 times the quotient of 72 and 8.

2 Answer: B

Explanation: Multiplying 40 x 10 x 10 results in 4,000. Multiplying 4 x 1 times 4. The given expression is 1,000 times 4 x 1.

3 Answer: $26,100

Explanation: If $27,000 is 30 times the initial deposit, find the initial deposit by dividing $27,000 by 30. The initial amount deposit is $900. The difference between this amount, and the total amount saved is $26,100.

4 Answer: No

Explanation: Four ten-thousands is 100 times larger than 4 hundreds. Each place value is 10 times the place value to the right. Since there are two positions between the hundreds and ten thousands place, the place value is multiplied by 100.

5 Answer: C

Explanation: The numerical representation of this situation is 17,002 − 3,019. The difference is 13,983.

6 Answer: 665 > 620

Explanation: Lara has $590 and earns $75, so she has $665. Karla has $470 and earns $150, so she has $620.

7 Answer: Eight million six hundred twenty-seven thousand six hundred twenty-eight

Explanation: Austria and Grenada are the most populated and least populated countries, respectively. Subtract their populations.

8 Answer: Six million three hundred seventy-seven thousand, eight hundred fifty-three.

Explanation: Express the number expressed in written form following the conventions of the base-ten number system.

9 Answer: Emil

Explanation: When counting by hundreds, the number 23,021 lies between 23,000 and 23,100, and is closest to 23,000.

10 Answer: A

Explanation: Rounding each number to the nearest thousand would result in 6,000 + 3,000 + 13,000 + 0 . The last number in the expression is less than 500, and rounded to the nearest thousand is 0.

11 Answer: 5,999 is closer to 6,000.

Explanation: The next hundred after 5,900 is 6,000. The number 5,999 is closest to 6,000. Not only does the hundreds place change, but the number of thousands increases as well.

12 Answer: A

Explanation: When adding the ones digit, the result is greater than 9, so the sum of the ones digit requires regrouping.

13 Answer: Students should describe a process which involves regrouping the ones and hundreds.

Explanation: Sample response: Add the ones: 5+9 = 14. Regroup the ones as 4 ones and 1 ten. Add the tens: 10 + 80 = 90 or 9 tens. Add the hundreds: 900 + 700 = 1,600. Regroup the hundreds as 6 hundreds and 1 thousand. Add the thousands: 5,000 + 1,000 = 6,000. Add all the partial sums: 6,000 + 600 + 90 + 4 = 6,694.

14 Answer: The original digits in the ones place should be 7 and 6. The original digits in the tens place could be 7 and 4, or 8 and 3. The original digits in the hundreds place could be 4 and 3, or 5 and 2.

Explanation: Sample Response:
477 + 346 = _____

15 Answer: 1,152

Explanation: Use the parts of the model to calculate the product: 40x20 + 8x20 + 40x4 + 8x4 = 1,152. The area model represents the expression 24x48.

16 Answer: 1,000 quarters

Explanation: Determine the total number of quarters by multiplying 25x40.

17 Answer: Mr. Ngo's class

Explanation: Mr. Ngo's class spends 31x$18 = $558 on the field trip, and Ms. Torres' class spends 27x$20 = $540.

18 Answer: 457

Explanation: There are 457 people in each section; from 2,742 divided by 6.

19 Answer: A

Explanation: Divide 928 by 6 using the standard algorithm or another strategy. The quotient is 154 with a remainder of 4.

20 Answer: Yes

Explanation: Jamal used the partial quotient strategy to divide the numbers. He Decomposed the number into three partial quotients that are divisible by 3 to divide the number.

NUMBER & OPERATIONS IN BASE TEN EXTRA PRACTICE

1 Answer: 800

Explanation: Pin City's population is 100 times smaller than the population of Mora. The digit 8 in the number 80,000 moves 2 positions to the right, leaving the value of 800.

2 Answer: N = 30
L = 300
K = 200
M = 2

Explanation: The value of each place-value position as 10 times the position to the right and as one-tenth of the value of the place to its left.

3 Answer: B

Explanation: Express the answer by dividing 5500 by 11.

4 Answer: 1,700

Explanation: She made 170 bracelets between January and May. Ten times this amount is 1,700.

5 Answer: D

Explanation: To write the inequality used to compare the values, find the difference between the amount of money Mr. Williams has and the cost of each car.

6 Answer: 276,543

Explanation: The number is 276,543. The number should have 6 digits, ending in the hundred thousands place. The digit in the ones place is one half of 6, or 3. Find the other digits in the number by adding 1 to the digit to the right.

7 Answer: The digit 3 is 10 times the digit 3 in the second number.

Explanation: Express the digit 3 as 3x1,000 in the first number, and the digit 3 as 3x100 in the second number.

8 Answer: 804,520 < 1,432,001

Explanation: The second number has a digit in the millions place, where the first number only has digits up to the hundred thousands place. The second number is larger.

9 Answer: A

Explanation: The value of the first set of coins rounds to $3700. The value of the second set of coins rounds to $1,200. The value of the third set of coins rounds to $1,000. Add the numbers.

10 Answer: 160

Explanation: The bar graph scale is marked by 10s, which makes identifying how close /far away each value is to the nearest 10 visual. Henry's collection is closest to 30, Jackson's collection is closest to 50, Mary's collection is closest to 30, and Xavier's collection is closest to 50. The total is 160.

11 Answer: 34,827

Explanation: Using the clues, the number rounds to 35,000 when rounded to the nearest thousand (meaning it could be in the 34 thousand or 35 thousand range). If the digit in the ones place is a 7, then the digits in the ten thousands and thousand places would have to be 3 and 4 respectively. The digit in the hundreds place, which is 1 more than the digit in the thousands place, must be 8.

12 Answer: No

Explanation: Evelyn subtracts smaller digits from the larger digits in each position without regrouping.

13 Answer: Yes

Explanation: The number of ones and tens in the minuend are less than the number of ones and tens in the subtrahend.

14 Answer: Yes

Explanation: Ben regroups when necessary when finding the difference in the numbers. Regrouping is necessary for the tens, hundreds, thousands, and ten thousands

15 Answer: D

Explanation: There are 3,129 people attending on Monday. There are (3,129x2) people attending for the next 4 days.

16 Answer: 896

Explanation: Multiply the dimensions of each part of the figure and add the areas: 50x10+50x6+6x10+6x6=896. The area model represents the expression 16x56.

17 Answer: Both expressions involve correct decomposition of the factors.

Explanation: Both strategies are correct Strategy A uses doubling, and then decomposes the largest factor (176) into ones, tens, and hundreds. Strategy B decomposes 88 into ones and tens.

18 Answer: 324/2=162
162/9=18

Explanation: Half of 324 is 162; there are 162 people remaining in the competition. If these people are divided into 9 equal groups, there will be 18 people in each group.

19 Answer: $546

Explanation: There is $840 from selling baseball bats. This means that $2226−$840 gives $1386, the money from selling towels. The difference between 1386 and 840 is 546. Alternatively, there are 105 baseball bats and 198 towels. Timothy's team earns $1386 from selling towels (198x7). The difference between $1386 and $840 is $546.

20 Answer: 27 bags

Explanation: The first step is to divide each original amount by the number of materials in each bag. Calculate 275 / 10 = 27 bags with 5 extra pencils, 180 erasers / 6 = 30 bags, and 328 / 9 = 36 bags with 4 extra pens. The number of bags is limited by the number of pens.

NUMBER & OPERATIONS - FRACTIONS UNIT 1: EQUIVALENT FRACTIONS AND ORDERING FRACTIONS

1 Answer: D

Explanation: The fraction model shows 1 part shaded out of 3 parts or $\frac{1}{3}$ of the entire figure shaded. An equivalent model shows 2 parts shaded out of 6 parts or $\frac{2}{6}$ shaded. The fraction $\frac{2}{6}$ is equivalent to $\frac{1}{3}$ because $\frac{1}{3}\times\frac{2}{2}=\frac{2}{6}$, and $\frac{2}{2}$ is equal to 1.

2 Answer: C

Explanation: The fraction $\frac{40}{100}$ is equivalent to $\frac{4}{10}$ because $\frac{4}{10}\times\frac{10}{10}=\frac{40}{100}$, and $\frac{10}{10}$ is equal to 1.

3 Answer: C

Explanation: The bar model shows 2 out of 4 parts shaded or $\frac{2}{4}$ of the figure shaded. The equivalent fraction to $\frac{2}{4}$ is $\frac{1}{2}$ because $\frac{1}{2}\times\frac{2}{2}=\frac{2}{4}$, and $\frac{2}{2}$ is equal to 1.

4 Answer: B

Explanation: The part is the not shaded on the bar model is 6 parts out of 8 parts, or $\frac{6}{8}$. The fraction $\frac{3}{4}$ is equivalent to $\frac{6}{8}$ because $\frac{3}{4}\times\frac{2}{2}=\frac{6}{8}$, and $\frac{2}{2}$ is equal to 1.

5 Answer: A

Explanation: The fraction of the stars that are red is 4 stars out of 10 stars or 4/10. The fraction $\frac{2}{5}$ is equivalent to $\frac{4}{10}$ because $\frac{2}{5}\times\frac{2}{2}=\frac{4}{10}$, and $\frac{2}{2}$ is equal to 1.

6 Answer: C

Explanation: The point on the number line is on $\frac{6}{9}$. $\frac{2}{3}$ is equivalent to $\frac{6}{9}$ because $\frac{2}{3}\times\frac{3}{3}=\frac{6}{9}$, and $\frac{3}{3}$ is equal to 1.

7 Answer:

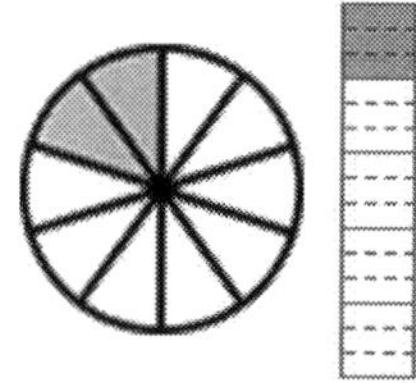

Explanation: Models should represent equal parts with so the ratio of the shaded parts to the unshaded parts is equivalent to the fraction $\frac{1}{5}$.

8 Answer: $\frac{3}{5}$, $\frac{12}{20}$, $\frac{18}{30}$

Explanation: The shaded portion represents the fraction $\frac{6}{10}$ which can be rewritten as the fraction $\frac{3}{5}$. The fractions $\frac{3}{5}$, $\frac{12}{20}$, and $\frac{18}{30}$ are equivalent to $\frac{6}{10}$.

9 Answer: B

Explanation: Of the chocolates, 4 out of 18 are white chocolate are the rest are milk chocolate. Subtract $18-4=14$ to find the number of milk chocolates. $18-4=14$. 14 out of 18 are milk chocolate $\frac{14}{18}$. $\frac{14}{18}\div\frac{2}{2}=\frac{7}{9}$.

10 Answer: B

Explanation: Shade 8 of the 12 parts in the figure. Notice that the ratio of shaded parts to the total area is $\frac{8}{12}$. This fraction is equivalent to $\frac{2}{3}$ because $\frac{2}{3}\times\frac{4}{4}=\frac{8}{12}$.

11 Answer: C

Explanation: Of $\frac{3}{10}$, $\frac{1}{3}$, $\frac{1}{5}$ and $\frac{3}{4}$, the smallest fraction is $\frac{1}{5}$. When comparing the fractions with a numerator of 1, the larger denominator indicates a smaller size fractional part. Alternatively, compare the decimal values of each fraction: $\frac{3}{10}=0.30$, $\frac{1}{3}=0.33$, $\frac{1}{5}=0.20$ and $\frac{3}{4}=0.75$.

12 Answer: D

Explanation: The fraction $\frac{11}{12}$ is the largest fraction because it is the closest to one whole unit.

13 Answer: B

Explanation: The fraction $\frac{1}{3}$ can be represented as $\frac{2}{6}$. $\frac{2}{6}<\frac{4}{6}$ or $\frac{4}{6}>\frac{2}{6}$.

14 Answer: B

Explanation: The fraction $\frac{2}{4}=\frac{1}{2}$, and $\frac{1}{2}\times\frac{2}{2}=\frac{2}{4}$, so $\frac{2}{4}$ and $\frac{1}{2}$ are equivalent or represent the same ratio.

15 Answer: A

Explanation: The fraction $\frac{1}{4}$ is less than $\frac{1}{2}$, $\frac{3}{6}$ is equal to $\frac{1}{2}$. Therefore $\frac{1}{4}<\frac{3}{6}$.

16 Answer: C

Explanation: The fraction $\frac{2}{4}$ is equal to $\frac{1}{2}$. Therefore, $\frac{2}{4}$ is less than any fraction in which the number in the denominator (bottom) is less than two times the number in the numerator (top) of the other fraction.

17 Answer: $\frac{4}{20}$, $\frac{2}{5}$, $\frac{60}{100}$, $\frac{20}{25}$

Explanation: The fractions can be compared as fractions out of 100: $\frac{4}{20}=20/100$; $\frac{2}{5}=40/100$; $\frac{20}{25}+=80/100$.

18 Answer: Greater than $\frac{1}{2}$: $\frac{6}{9}$, $\frac{7}{10}$, $\frac{4}{5}$
Equivalent to $\frac{1}{2}$: $\frac{3}{6}$, $\frac{7}{14}$, $\frac{50}{100}$

Explanation: A fraction is greater than $\frac{1}{2}$ if the numerator is more than half of the denominator. A fraction is equivalent to $\frac{1}{2}$ if the numerator is exactly half of the denominator.

19 Answer: C

Explanation: The fraction $\frac{3}{4}$ was the largest amount of time studied because $\frac{3}{4}$ is the largest fraction listed.

20 Answer: A

Explanation: As decimals, Sam's fraction $\frac{6}{8}$ is 0.75, Emma's fraction $\frac{4}{5}$ is 0.80, and Kevin's fraction $\frac{2}{4}$ is 0.50. The value of the fraction $\frac{6}{8}$ is in between the values of $\frac{2}{4}$ and $\frac{4}{5}$. The fraction $\frac{4}{5}$ is the largest fraction, the fraction $\frac{6}{8}$ is in the middle, and $\frac{2}{4}$ is the smallest fraction.

NUMBER & OPERATIONS - FRACTIONS
UNIT 2: FRACTION OPERATIONS

1 Answer: C

Explanation: The first model has 2 parts shaded out of 6 parts, which is $\frac{2}{6}$. The second model has 4 parts shaded out of 6 parts, which is $\frac{4}{6}$. The model adds the two fractions: $\frac{2}{6}+\frac{4}{6}$.

2 Answer: B

Explanation: When adding fractions with the same denominator, add the numerators, but do not change the denominators. ($\frac{1}{5}+\frac{1}{5}+\frac{1}{5}+\frac{1}{5}=\frac{4}{5}$)

3 Answer: A

Explanation: Add the numerators. ($\frac{2}{8}+\frac{5}{8}=\frac{7}{8}$) Do not change the denominators.

4 Answer: D

Explanation: Subtract the numerators, but do not change the denominators. ($\frac{9}{10}-\frac{5}{10}=\frac{4}{10}$)

5 Answer: C

Explanation: The shaded part of the model is 4 parts out of 10 so $\frac{4}{10}$ is shaded. This region can be written in four parts: $\frac{1}{10}+\frac{1}{10}+\frac{1}{10}+\frac{1}{10}=\frac{4}{10}$.

6 Answer: D

Explanation: Adding the numerators, $\frac{7}{12}+\frac{6}{12}=\frac{13}{12}$, shows that the resulting fraction has a numerator that is larger than the denominator, which means the fraction is larger than 1.

7 Answer: B

Explanation: A whole pizza cut into 8 slices is $\frac{8}{8}$. If Dane ate $\frac{4}{8}$ of the pizza, subtract to find the remaining part: $\frac{8}{8}-\frac{4}{8}=\frac{4}{8}$.

8 Answer: A

Explanation: Change the mixed numbers into improper fractions. First rewrite 3 as a fraction with the same denominator as the fraction within the mixed number. Then, add $\frac{12}{4}+\frac{3}{4}=\frac{15}{4}$. Next rewrite 2 as a fraction with the same denominator. Then, add $\frac{8}{4}+\frac{2}{4}=\frac{10}{4}$. Add the results.

9 Answer: D

Explanation: The mixed number $3\frac{2}{5}$ can be rewritten as an improper fraction of $\frac{17}{5}$. This fraction contains 17 groups of $\frac{1}{5}$. The mixed number $2\frac{4}{5}$ can be rewritten as an improper fraction $\frac{14}{5}$. This fraction contains 14 groups of $\frac{1}{5}$. Add 17 and 14.

10 Answer: C

Explanation: The mixed number $7\frac{3}{4}$ can be rewritten as an improper fraction of $\frac{31}{4}$ which contains 31 groups of $\frac{1}{4}$. This means Karl has practiced $\frac{1}{4}$ hours for 31 days.

11 Answer: B

Explanation: The fraction $\frac{2}{5}$ of 15 is 6 people. This means 6 people out of 15 people brought gifts, and 9 people out of 15 people did not bring gifts.

12 Answer: C

Explanation: If $\frac{3}{4}$ of the students are going to the fair, and $\frac{1}{3}$ of those students ride the bus, then the portion of the students who rode the bus to the fair is $\frac{3}{4}\times\frac{1}{3}=\frac{1}{4}$. The group shows 12 people; $\frac{3}{4}$ of 12 is 9, and $\frac{1}{3}$ of 9 is 3.

13 Answer: $3\frac{1}{3}$ or $\frac{10}{3}$

Explanation: The amount of ribbon needed to wrap one gift is $\frac{2}{3}$ feet in length. Multiply $\frac{2}{3}$ by 5 and get $\frac{10}{3}$ feet or $3\frac{1}{3}$ feet.

14 Answer: A

Explanation: Using the given information, $\frac{4}{8}$ of the pencils are yellow, and since there are 32 pencils, this is half of the original amount. This means 16 pencils were yellow.

15 Answer: B

Explanation: There are 8 measuring cups that are filled to the $\frac{5}{8}$ mark. The total amount of peanuts can be represented with this expression: $\frac{5}{8} \times 8$.

16 Answer: 15

Explanation: Find the answer by multiplying 3×5.

17 Answer: B

Explanation: He needs $1\frac{1}{3}$ cups of liquid for one serving. To make 5 servings, multiply 5 by $1\frac{1}{3}$ cups. He will need $6\frac{2}{3}$ cups.

18 Answer: D

Explanation: Multiply $\frac{2}{3}$ by 18, which results in 12 bananas. Dividing 18 by $3 = 6$. This represents $\frac{1}{3}$ of the bananas. Then, $\frac{1}{3} + \frac{1}{3} = \frac{2}{3}$; this means $6 + 6 = 12$.

19 Answer: $4 \times \frac{1}{3}$

Explanation: There are 4 squares with 1/3 of each square shaded.

20 Answer:

$\frac{35}{8}$ or $4\frac{3}{8}$ cups of fruit

$\frac{21}{8}$ or $2\frac{5}{8}$ cups of juice

Explanation: The given recipe is for 2 servings, and to determine the amount of fruit and juice needed for 14 servings. Multiply each amount by 7.

NUMBER & OPERATIONS - FRACTIONS UNIT 3: DECIMAL CONVERSION AND COMPARISON

1 Answer: B

Explanation: The length of the leaf is $3\frac{2}{10}$ cm. The fraction $\frac{2}{10}$ is equivalent to $\frac{20}{100}$.

2 Answer: C

Explanation: The units on the number line are divided by 10. Point A is on the 3rd mark. The fraction $\frac{3}{10}$ is equivalent to $\frac{30}{100}$: $\frac{3}{10} \times (\frac{10}{10}) = \frac{30}{100}$

3 Answer: $\frac{10}{10}$ or $\frac{100}{100}$ or 1

Explanation: The fraction in Model A is $\frac{9}{10}$. The fraction in Model B is $\frac{10}{100}$. The sum of $\frac{9}{10}$ and $\frac{10}{100}$ is the same as $\frac{9}{10} + \frac{1}{10} = \frac{10}{10} = 1$.

4 Answer: No

Explanation: Each mark on the number line represents $\frac{1}{100}$. The difference between Point X and Point Y is $\frac{4}{100}$.

5 Answer: $\frac{5}{10}$ or $\frac{50}{100}$

Explanation: The fraction in Model A is $\frac{3}{10}$. The fraction in Model B is $\frac{20}{100}$.
Add $\frac{3}{10}$ and $\frac{20}{100}$ as $\frac{3}{10} + \frac{2}{10}$. The result is $\frac{5}{10}$.

6 Answer: A

Explanation: Model A shows 7 bars shaded out of 10. This represents $\frac{7}{10}$. Model B shows 70 squares shaded out of 100. This represents $\frac{70}{100}$. The fraction $\frac{70}{100}$ is equivalent to the fraction $\frac{7}{10}$.

7 Answer: B

Explanation: Multiplying a fraction by 1 (represented as a fraction $\frac{a}{a}$) creates an equivalent fraction. Thus, multiplying the fraction $\frac{5}{10}$ by $\frac{10}{10}$ does not change its value.

8 Answer: B

Explanation: The fraction $\frac{4}{10}$, read as "four tenths", is expressed as the decimal 0.4.

9 Answer: D

Explanation: The mixed number contains the fraction $\frac{3}{10}$. Thus, consider that the mixed number is $6+\frac{3}{10}$. Since the fraction $\frac{3}{10}$ equals the decimal 0.3. The mixed number equals $6+0.3$ which means the mixed number is equal to the decimal 6.3.

10 Answer: C

Explanation: The scissors are $\frac{6}{10}$ ft. long. To convert this fraction to a decimal, divide 6 by 10.

11 Answer: D

Explanation: The fraction $\frac{74}{100}$ is equivalent to the decimal 0.74. Since this is a mixed number, the 32 is a whole number plus the decimal and is represented as 32.74.

12 Answer: 0.19

Explanation: The sum of the fractions given in the table are or $\frac{81}{100}$, when expressed as a decimal is 0.81. Subtracting this value from 1 whole gives 0.19.

13 Answer: Yes

Explanation: The difference between Point A and Point C is $\frac{2}{10}$ or 0.2.

14 Answer: 0.22

Explanation: Evie's model has a shaded value of $\frac{71}{100}$ or 0.71 and Parth's model has a shaded value of $\frac{49}{100}$ or 0.49. The difference is 0.22.

15 Answer: B

Explanation: The fraction $\frac{1}{2}$ is 0.5 or 0.50 as a decimal, and the fraction $\frac{3}{4}$ is 0.75 as a decimal. Both decimals are greater than 0.5 and less than 0.75.

16 Answer: C

Explanation: Lulu has \$0.42 and Mario and Leo together have \$0.58. The value of \$0.58 is greater than the value of \$0.42.

17 Answer: C

Explanation: The fraction $\frac{1}{4}$ is 0.25 or 0.250, and the fraction $\frac{3}{8}$ is 0.375. Both decimals are between $\frac{1}{4}$ and $\frac{3}{8}$.

18 Answer: B

Explanation: The fraction $\frac{1}{4}$, as a decimal, is 0.25, and $\frac{3}{5}$, as a decimal, is 0.6. The distance between each mark on the number line is 0.1. Point B represents the decimal 0.3. Point C represents the decimal 0.4. Point D represents the decimal 0.5.

19 Answer: C

Explanation: The distance between each hash mark on the number line is 0.01. Point x represents the decimal 0.82.

20 Answer: $0.65 < 0.75$ or $0.75 > 0.65$

Explanation: The distance between each hash mark is 0.01. Point A is located at 0.65. Point B is located at 0.75.

NUMBER & OPERATIONS - FRACTIONS CHAPTER REVIEW

1 Answer: A

Explanation: The fraction $\frac{4}{32}$ can be simplified into an equivalent fraction $\frac{1}{8}$ because 4 in the numerator and 32 in the denominator have a common factor of 4 which can be divided out: $\frac{4}{32} \div \frac{4}{4} = \frac{1}{8}$.

2 Answer: C

Explanation: The fraction $\frac{1}{4}$ is equivalent to $\frac{2}{8}$ because $\frac{1}{4} \times \frac{2}{2} = \frac{2}{8}$.

3 Answer: C

Explanation: The fraction $\frac{2}{3}$ is larger than $\frac{1}{2}$. The denominator is less than two times the numerator.

4 Answer: $\frac{5}{8} > \frac{1}{2}$

Explanation: The fraction is $\frac{5}{8}$ is larger than $\frac{1}{2}$ because $\frac{1}{2} = \frac{4}{8}$. Use the given figures to show the inequality by shading 5 parts in the first figure and shading 4 parts in the second figure.

5 Answer: B

Explanation: The total amount of flour is $\frac{1}{6} + \frac{4}{6}$ or $\frac{5}{6}$. The fraction $\frac{5}{6}$ is equivalent to $\frac{1}{6}$ added to itself 5 times.

6 Answer: C

Explanation: Represent the original amount of soda with the fraction $\frac{8}{8}$. The difference between $\frac{5}{8}$ and $\frac{8}{8}$ is $\frac{3}{8}$. Then, $\frac{3}{8}$ is equal to $\frac{1}{8}$ added to itself 3 times.

7 Answer: D

Explanation: Multiply $\frac{1}{3}$ by 12 or divide 12 by 3 to get 4. Multiply $\frac{1}{4}$ by 24 or divide 24 by 4 to get 6.

8 Answer: $6 \times \frac{1}{4}$
$\frac{6}{4}$ or $1\frac{2}{4}$

Explanation: There are 6 squares with $\frac{1}{4}$ of each square shaded. Multiply 6 by $\frac{1}{4}$ which results in $\frac{6}{4}$ or $\frac{3}{2}$ or $1\frac{1}{2}$.

9 Answer: D

Explanation: Multiply 40 by 10 and multiply $\frac{5}{100}$ by $\frac{10}{1}$ to get $\frac{50}{100}$. Combine the two results into a new mixed number.

10 Answer: D

Explanation: Point R is located at $\frac{2}{10}$ and Point S is located at $\frac{5}{10}$. The sum is $\frac{7}{10}$. Multiplying $\frac{7}{10}$ by $\frac{10}{10}$ (or 1) gives the fraction $\frac{70}{100}$.

11 Answer: $\frac{60}{100}$

Explanation: To have an equivalent fraction with a denominator of 100, multiply the numerator and denominator by 10.

12 Answer: $\frac{9}{10}$ or $\frac{90}{100}$

Explanation: Point X is located at $\frac{8}{10}$, which is equivalent to $\frac{80}{100}$. Adding $\frac{10}{100}$ to this fraction is $\frac{90}{100}$ or $\frac{9}{10}$.

13 Answer: D

Explanation: Adding $\frac{8}{10}$ to $9\frac{3}{10}$ is the same as adding 0.8 to 9.3. The result is 10.1.

14 Answer: $\frac{35}{100}+\frac{5}{10}=0.85$

Explanation: The two fractions add up to 0.85. The fraction $\frac{35}{100}=0.35$ and the fraction $\frac{5}{10}=0.50$. The students have completed 0.85 of the book report.

15 Answer: 0.3 or 0.30

Explanation: The fraction $\frac{15}{100}$ is 0.15 as a decimal. Adding 0.15+ 0.15 is equivalent to the fraction of the students who prefer green.

16 Answer: 0.34

Explanation: The figure shows that 34 squares out of 100 squares are shaded. The shaded portion is equivalent to $\frac{34}{100}$ or 0.34.

17 Answer: <

Explanation: The decimal 1.85 is less than 1.9 because 1.9 can be written as 1.90.

18 Answer: A

Explanation: Umar has $0.68, Zane has $0.79, and $0.79 is greater than $0.68.

19 Answer: D

Explanation: Zack has $0.78, Lara has $0.14, and $0.78 is greater than $0.14

20 Answer: 18.4 < 24.9

Explanation: The decimal 18.4 is less than the decimal 24.9.

NUMBER & OPERATIONS - FRACTIONS EXTRA PRACTICE

1 Answer: $\frac{2}{3}; \frac{8}{12}; \frac{12}{18}$

Explanation: The "X" is located at $\frac{4}{6}$.

2 Answer: 1

Explanation: 5 out of the 30 shells are black so $\frac{5}{30}$ are black. If she adds 6 shells with the same fractional amount being black, find the number of black shells with the proportion $\frac{5}{30}=\frac{x}{6}$. Solve this problem using multiplication or division. $6 \times 5 = 30x$, so $x = 1$. The unknown numerator is 1 so 1 out of the 6 shells are black.

3 Answer: A

Explanation: The fraction $\frac{1}{4}$ is halfway between 0 and $\frac{1}{2}$. The fraction $\frac{3}{7}$ is closer to $\frac{1}{2}$. Therefore, $\frac{3}{7}$ is the larger fraction.

4 Answer: Yes, he spent more than $\frac{3}{5}$ of an hour on his homework

Explanation: Compare the two fractions with the inequality $\frac{3}{5}<\frac{4}{6}$.

5 Answer: Yes

Explanation: The number line shows 4 jumps which represent the 4 study times. The distance between hash marks is $\frac{1}{4}$ of an hour. Lisa studies for $\frac{9}{4}$ or $2\frac{1}{4}$ hours.

6 Answer: C

Explanation: Subtract $\frac{2}{10}$ and $\frac{1}{10}$ from $\frac{10}{10}$. The children ate $\frac{7}{10}$ of the pizza.

7 Answer: $\frac{1}{5}+\frac{4}{5}+\frac{3}{5}=\frac{8}{5}$
$\frac{8}{5} \times 7=\frac{56}{5}$
$\frac{56}{5}=11\frac{1}{5}$

Explanation: John walks one-fifth miles each day; Tyler walks four-fifth miles each day; Erick walks three-fifth miles each day. Over a seven-day period, find the total number of miles they walk by multiplying their daily distance by 7.

8 Answer: Waymon's model is correct. Tessa' model is incorrect.

Explanation: Waymon's model shows 3 squares, with each square divided into 2 equal parts, creating 6 halves. Five of the halves are shaded. This represents $5 \times \frac{1}{2}$. Tessa's model shows 5 rectangles each divided into 2 unequal parts. The parts do not represent halves.

9 Answer: $\frac{10}{100} + \frac{40}{100}$ **or** $\frac{1}{10} + \frac{4}{10}$

Explanation: Convert the fractions to a common denominator before adding: $\frac{10}{100}$ is equivalent $\frac{1}{10}$ and $\frac{4}{10}$ is equivalent to $\frac{40}{100}$.

10 Answer: Yes

Explanation: The fraction $\frac{80}{100}$ is equivalent to $\frac{8}{10}$, and there are 6 pennies.

11 Answer: B

Explanation: Model B has 80 squares out of 100 squares shaded. This fraction is $\frac{80}{100}$ or $\frac{8}{10}$. In Model A, 8 of the 10 parts should be shaded.

12 Answer: A

Explanation: To find the difference subtract $56 - 8 = 46$; $\frac{7}{10}$ is equivalent to $\frac{70}{100}$.

13 Answer: Divide 87 by 10. Write the quotient in decimal form.

Explanation: The fraction given in the stem is an improper fraction. When converted, the decimal form of this fraction is 8.70 or 8.7. Dividing by 10 moves the decimal of the numerator one place to the left.

14 Answer: No

Explanation: The distance between each hash mark is 0.1. The difference between Point A and Point C is 1.3.

15 Answer: A

Explanation: The ingredients take up $\frac{8}{10}$ or 0.8 of the pot. To find the remaining space, subtract $\frac{8}{10}$ from $\frac{10}{10}$, then convert to a decimal.

16 Answer: B

Explanation: To convert $\frac{54}{100}$ into a decimal, divide the numerator by the denominator by moving the decimal in the numerator 2 places to the left. The number. 54 divided by 100 is 0.54.

17 Answer: 0.84

Explanation: There are 8 squares shaded, which represent 0.08. The difference between 0.92 and 0.08 is 0.84.

18 Answer: 1.33

Explanation: The smaller decimal is 0.37. Add 0.37 and 0.96. The result is 1.33.

19 Answer: $\mathbf{1.84 > 1.66}$ $\mathbf{1.66 < 1.84}$

Explanation: Nate's plant is shorter by 0.18 feet, which makes it 1.66 feet tall.

20 Answer: No

Explanation: Although 0.50 is greater than 0.25, the size of Winnie's popcorn bucket is larger than Michael's bucket. Based on the size of each bucket, $\frac{1}{2}$ of Michael's bucket is less than $\frac{1}{4}$ of Winnie's bucket.

MEASUREMENT & DATA
UNIT 1: PROBLEM SOLVING - CONVERSION OF MEASUREMENTS

1 Answer: A

Explanation: The length of 35 inches is the best estimate. All other lengths are not realistic.

2 Answer: A

Explanation: A normal telephone pole is 45-65 feet tall. Thus, 18 yards is a reasonable estimate.

3 Answer: A

Explanation: One yard is equivalent to 3 feet or 36 inches.

4 Answer: A

Explanation: One ton = 2,000 pounds and 2,000 ounces = 125 pounds.

5 Answer: C

Explanation: One gallon = 4 quarts = 8 pints.

6 Answer: B

Explanation: 1 pint 0.5 quarts = 2 cups.

7 Answer: D

Explanation: Pint are the largest unit measurement and has the greatest volume in this situation.

8 Answer: A

Explanation: $\$8.55-\$8.07=\$0.48$

9 Answer: B

Explanation: Subtract \$2.73 from \$7.50 to find the amount of the change.

10 Answer: C

Explanation: Subtract \$3.77 from \$6.25 to find the amount of the change.

11 Answer: B

Explanation: Subtract \$3.64 from \$5.00 to find the amount of the change.

12 Answer: D

Explanation: Subtract \$5.24 from \$5.50 to find the amount of the change.

13 Answer: A

Explanation: From 9AM to noon is 3 hours. Then, from noon to 2:30PM is 2 hours 30 minutes. Add the two quantities of time together.

14 Answer: A

Explanation: Five gallons = 20 quarts = 40 pints = 80 cups = 640 ounces.

15 Answer: C

Explanation: Find the perimeter by adding lengths of the 4 sides: $4+5+4+5=18$.

16 Answer: A

Explanation: Find the perimeter by adding the lengths of the 4 sides: $1+2+1+2=6$.

17 Answer: B

Explanation: Find the perimeter by adding the lengths of the 4 sides: $2+4+2+4=12$.

18 Answer: C

Explanation: Find the perimeter by adding the lengths of the 4 sides: $7+12+7+12=38$.

19 Answer: D

Explanation: Find the perimeter by adding the lengths of the 6 sides: $5+3+4+3+1+6=22$.

20 Answer: 8

Explanation: Find the perimeter by adding the lengths of the 6 sides: $1+2+2+1+1+1=8$.

MEASUREMENT & DATA
UNIT 2: GRAPHS AND DATA INTERPRETATION

1 Answer:

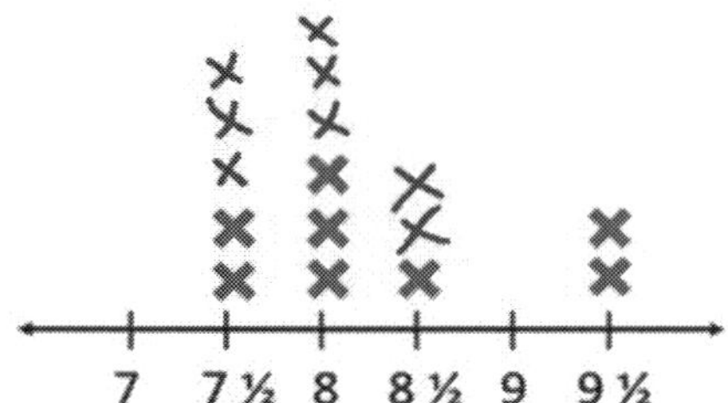

Explanation: Add 8 additional data points, from the table, to the line plot so the frequency of each value match.

2 Answer:

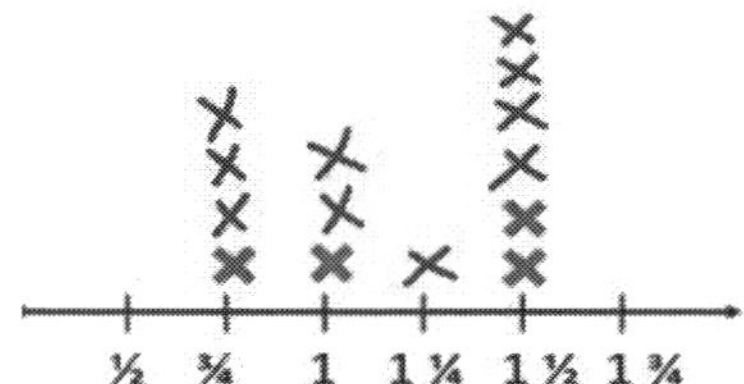

Number of Pies Made With Varying Cups of Fruit

Explanation: Add 10 additional data points, from the table, to the line plot so the frequency of each value match.

3

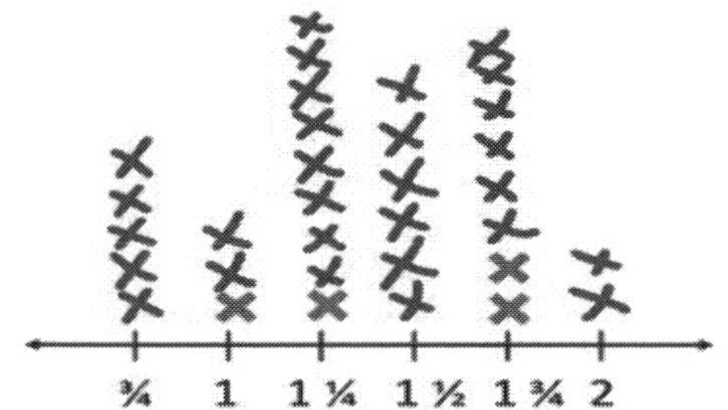

Number of Cookie Batches Baked

Explanation: Add 29 additional data points, from the table, to the line plot so the frequency of each value match.

4 Answer:

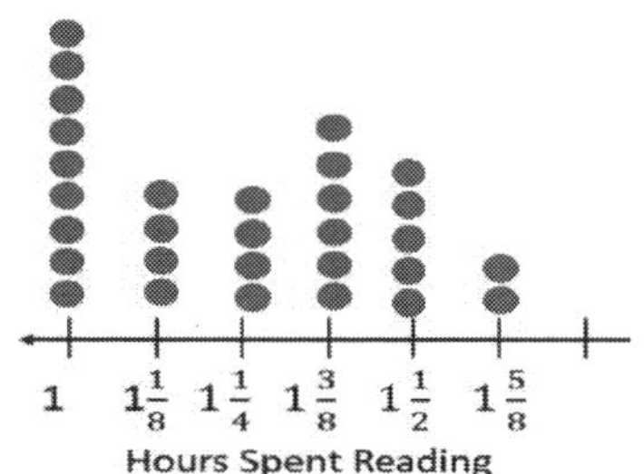

Explanation: Mark the 6 number categories on the line. Then, mark the tallies for each category above the appropriate value. Each tally mark in the table corresponds to a point on the line plot.

5 Answer:

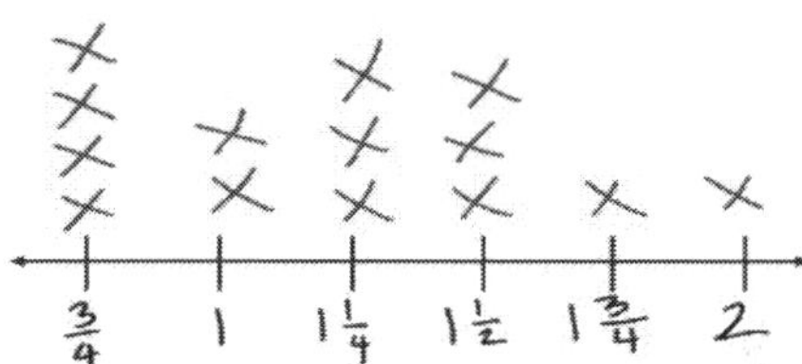

Explanation: Each value in the data table corresponds to a point on the line plot. The value in each row of the table is represented as a column in the line plot.

6 Answer:

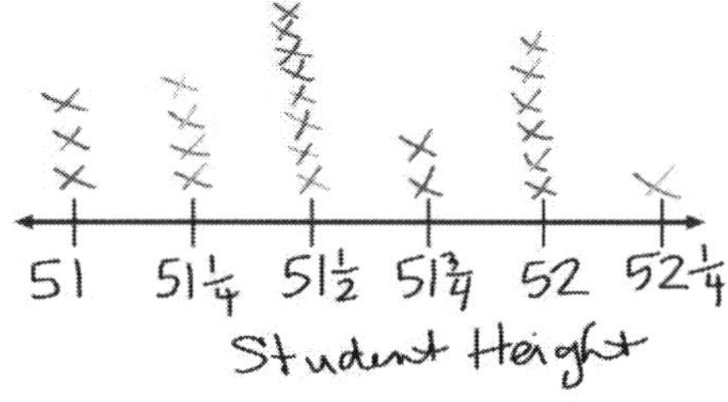

Explanation: Each value in the data table corresponds to a point on the line plot. The value in each row of the table is represented as a column in the line plot.

7 Answer: 9

Explanation: Determine the number of students by counting the marks to the left of the 1 1⁄4 cup position.

8 Answer:

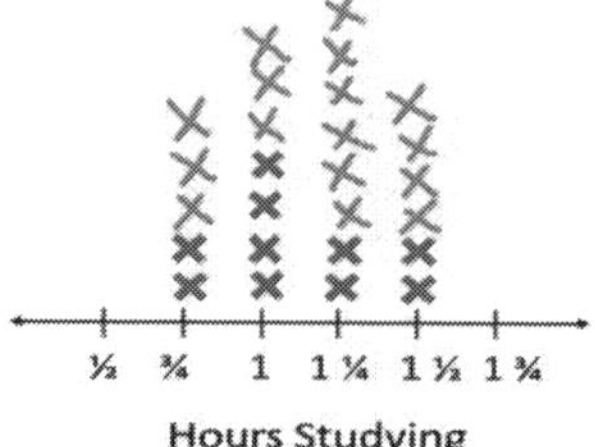

Explanation: Add 16 additional data points, from the table, to the line plot so the number of tallies match.

9 Answer:

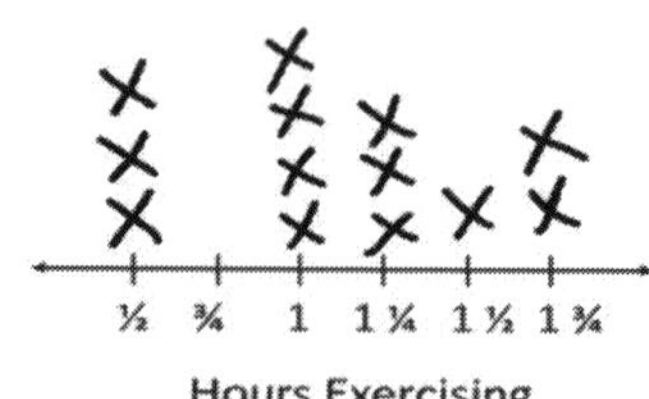

Explanation: The data set has 13 values that match the numbers on the number line. Plot the frequency of each value above the line in the appropriate place.

10 Answer:

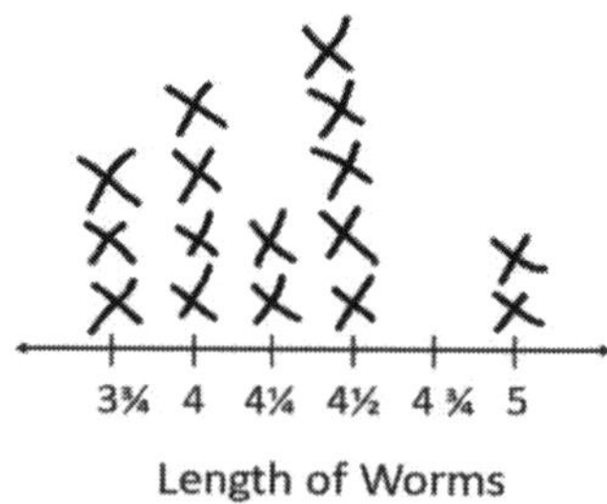

Explanation: The data set has 16 values that match the numbers on the number line. Plot the frequency of each value above the line in the appropriate place. The value 4 3⁄4 inches does not have any data points associated with it.

11 Answer:

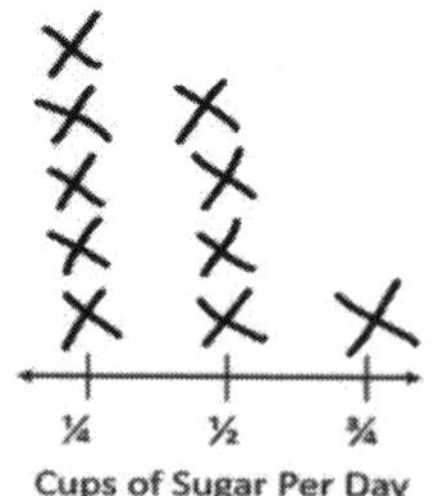

Explanation: There are 10 data points in this set. Each data point on the line plot represents the frequency of each value in the list.

12 Answer:

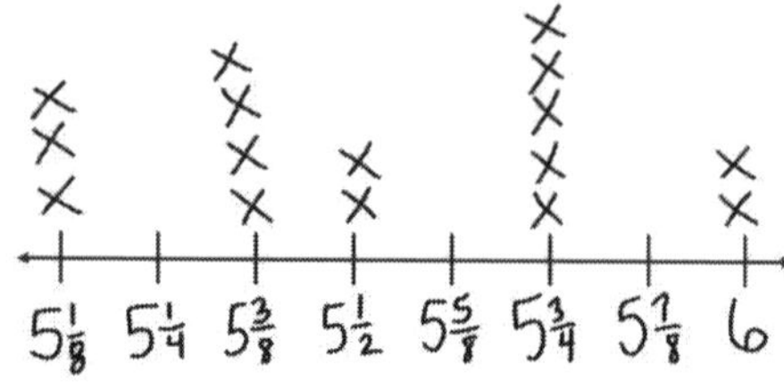

Explanation: Mark the 5 number categories on the line. Then, mark the tallies for each category above the appropriate value. Each tally mark in the table corresponds to a point on the line plot.

13 Answer:

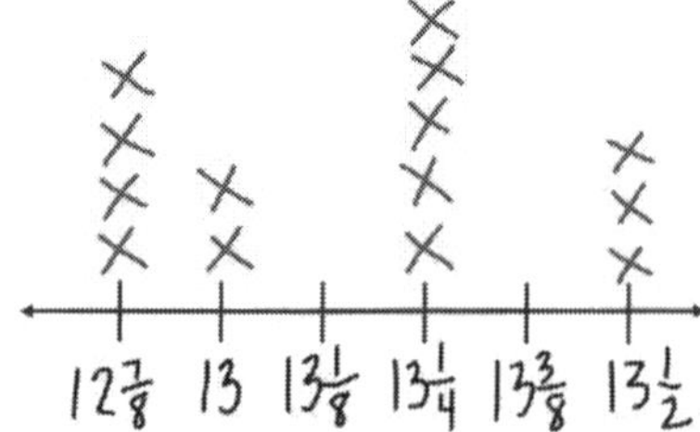

Explanation: Mark the 6 number categories on the line. Then, mark the tallies for each category above the appropriate value. Each tally mark in the table corresponds to a point on the line plot. There are 14 data points in the line plot. Some values do not have any data points associated with them.

14 Answer:

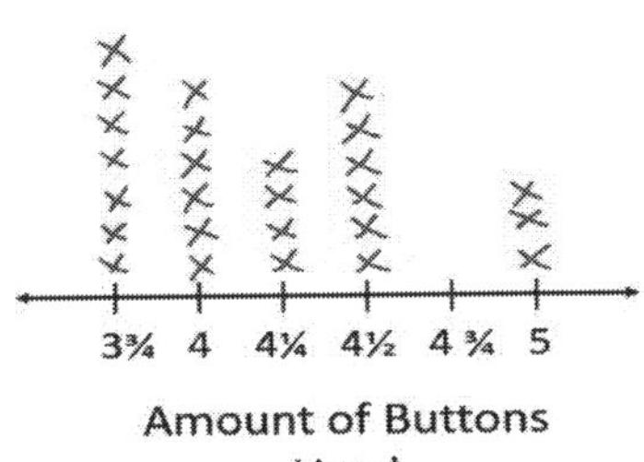

Explanation: There are 26 data points in the line plot. One value is included in the scale ($4\frac{3}{4}$) but will not have any data points associated with it.

15 Answer:

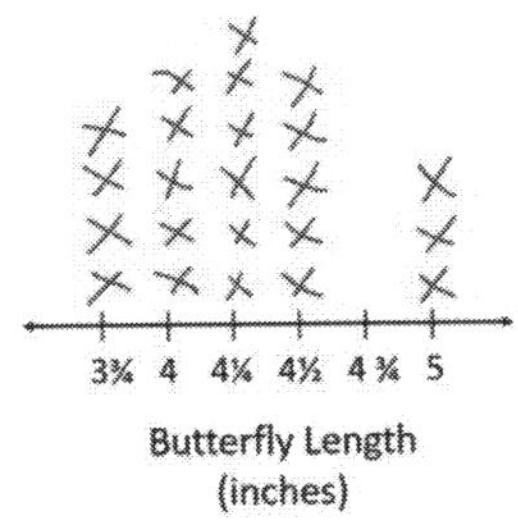

Explanation: There are 23 data points in the line plot. One value is included in the scale ($4\frac{3}{4}$) but will not have any data points associated with it.

16 Answer:

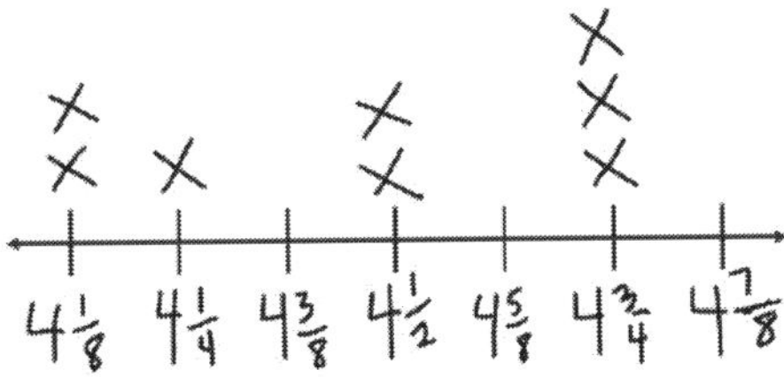

Explanation: There are 8 data points on the line plot. Some values are included in the scale but will not have any data points associated with them.

17 Answer:

Ounces of Cologne

$1 \quad 1\frac{1}{4} \quad 1\frac{1}{2} \quad 2$

Explanation: The frequency of each value determines the number of points at each value on the line plot. For example, the value "2" occurs 3 times in the data set. It is marked with 3 points.

18 Answer:

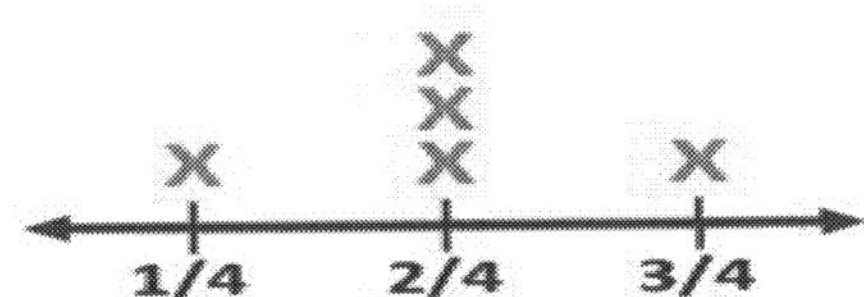

Explanation: The frequency of each value determines the number of points at each value on the line plot. For example, the value "2/4" occurs 3 times in the data set. It is marked with 3 points.

19 Answer:

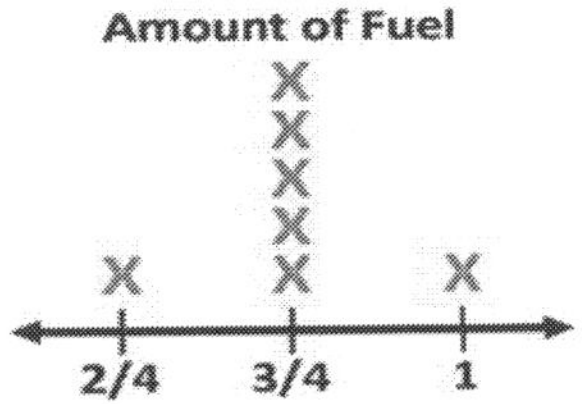

Explanation: The frequency of each value determines the number of points at each value on the line plot. For example, the value "3/4" occurs 5 times in the data set. It is marked with 5 points.

20 Answer:

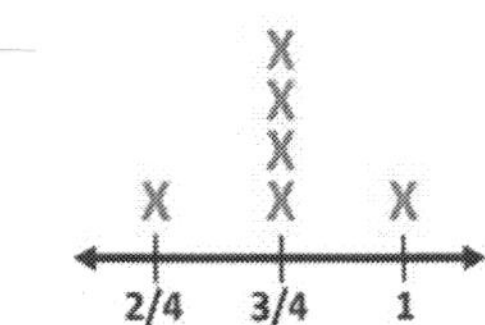

Explanation: The frequency of each value determines the number of points at each value on the line plot. For example, the value "3/4" occurs 4 times in the data set. It is marked with 4 points.

MEASUREMENT & DATA
UNIT 3: GEOMETRY - CONCEPTS OF ANGLES

1 Answer: A

Explanation: Based on the arc, the angle turns through 0 degrees of the 360 degrees in a circle. The angle has a measure of 0 degrees.

2 Answer: A

Explanation: Based on the shaded portion of the arc, the angle turns 180 degrees of the 360 degrees in a circle. The angle has a measure of 180 degrees.

3 Answer: D

Explanation: The angle is greater than a right angle (90 degrees) and less than a straight angle (180 degrees). The best estimate is 120 degrees.

4 Answer: $\frac{3}{4}$

Explanation: Based o the arc, the angle turns through 270 degrees out of 360 degrees in the circle. This fraction can be simplified to $\frac{3}{4}$.

5 Answer: 160

Explanation: Calculate the total movement by multiplying: $8 \times 20 = 160$.

6 Answer: 10

Explanation: Divide 600 by 60 to determine the number of degrees moved.

7 Answer: 3

Explanation: There are 360 degrees in a circle. Divide 360 by 120.

8 Answer: A

Explanation: Determine the angle measure by subtracting 0 from 60.

9 Answer: C

Explanation: Determine the angle measure by subtracting 0 from 90.

10 Answer: 120

Explanation: Determine the angle measure by subtracting 0 from 120.

11 Answer: D

Explanation: Determine the angle measure by subtracting 33 from 180.

12 Answer: C

Explanation: Determine the angle measure by subtracting 102 from 180.

13 Answer: 6

Explanation: Determine the angle measure by finding the difference between 96 and 90.

14 Answer: C

Explanation: Determine the angle measure by subtracting 0 from 157.

15 Answer: C

Explanation: Adjacent angles share a vertex and a side. Add the measures of adjacent angles to find the total measure: $40+30=70$.

16 Answer: 47.1

Explanation: The two angles are adjacent angles. Subtract 31.5 from 78.6.

17 Answer: B

Explanation: The angles are adjacent angles. Add the measures of adjacent angles to find the total measure: $30+50=80$.

18 Answer: 29

Explanation: The two angles are adjacent angles. To find the measure, subtract 49 from 78 degrees.

19 Answer: D

Explanation: The pizza is in the shape of a circle, which has 360 degrees. Each slice of pizza is 360/5, or 72 degrees. Dividing 72 by 2 gives the measure of half of the shared slice.

20 Answer: 28

Explanation: The two angles are adjacent angles. Find the measure of the angle by subtracting: $71-43=28$.

MEASUREMENT & DATA CHAPTER REVIEW

1 Answer: B

Explanation: One ton = 2,000 pounds (lbs.)

2 Answer: D

Explanation: Only 25 millimeters, approximately 1 inch, is a reasonable estimate.

3 Answer: B

Explanation: Only 8 liters, approximately 2 gallons, is a reasonable estimate.

4 Answer: A

Explanation: To find the unit price, divide the total cost by the number of packs.

5 Answer: A

Explanation: Subtract the number of quarts from each other (6–3) which leaves 3 quarts and 1 pint.

6 Answer: C

Explanation: She used 4 teaspoons of baking powder. This is equivalent of 1 and 1/3 tablespoons. Adding this amount to the 2 tablespoons of sugar means 3 1/3 tablespoons of sugar and baking powder are used.

7 Answer: C

Explanation: Find the perimeter of the figure by adding the lengths of the 6 sides: $3+2+2+3+1+5=16$.

8 Answer: B

Explanation: Find the perimeter of the figure by adding the lengths of the 6 sides: $1+2+2+1+3+3=12$.

9 Answer: B

Explanation: Find the perimeter of the figure by adding the lengths of the 6 sides: $7+2+4+2+3+4=22$.

10 Answer: $\frac{3}{4}$

Explanation: There are 5 grills which use ¾ gallons of fuel. This is the largest frequency shown on the line plot.

11 Answer: 3

Explanation: Count the number of data points to the right of $8\frac{1}{2}$ on the number line. There are 3 pencils longer than $8\frac{1}{2}$ inches.

12 Answer: D

Explanation: The angle is slightly less than a right angle (90 degrees) or 90/360 of the circle.

13 Answer: $\frac{1}{2}$

Explanation: This is a straight angle. It is a $\frac{1}{2}$ of a turn around a circle.

14 Answer: 360

Explanation: An angle that turns through an entire circle has a measure of 360 degrees.

15 Answer: 180

Explanation: An angle that turns through half a circle has a measure of 180 degrees.

16 Answer: B

Explanation: The angle has a measure of 45 degrees. An angle twice as large has a measure of 45x2, or 90 degrees.

17 Answer: No

Explanation: The measure of the angle is 90 degrees. Find the measure by subtracting 80 from 170.

18 Answer: D

Explanation: The angles are adjacent angles. Find the measure by subtracting: 80 − 30 = 50.

19 Answer: A

Explanation: The angles are adjacent angles. Find the measure by subtracting: 70 − 30 = 40.

20 Answer: 41

Explanation: The corner of the farm is a 90-degree angle. The two fences create a 15-degree and 34-degree angle (15 + 34 = 49), Subtract 49 from 90.

MEASUREMENT & DATA EXTRA PRACTICE

1 Answer: Missing values (from left to right): 6, 8, 6, 7, 16

Explanation: One pint = 2 cups. Use this conversion to complete the table. To convert from cups to pints, divide by 2. To convert from pints to cups, multiply by 2.

2 Answer: 1,915 millimeters or 1,915 mm

Explanation: One meter = 1,000 millimeters.

3 Answer: 1 kilometer or 1 km

Explanation: One kilometer = 1,000 meters.

4 Answer: A

Explanation: Add the times to find the total elapsed time. Next, find the time that is 2 hours and 55 minutes after 2:40 P.M.

5 Answer: B

Explanation: Add the minutes to find the total time per day. Next, multiply 35 by 5. Then, convert to hours and minutes.

6 Answer: B

Explanation: Add the times to find the total elapsed time.

7 Answer: C

Explanation: Add all the lengths of the 6 sides.

8 Answer: D

Explanation: Multiply the length times the height: 9 × 6 = 54.

9 Answer: A

Explanation: Multiply the side length by itself: 5 × 5 = 25.

10 Answer: 1

Explanation: Count the number of data points to the left of 3/4. One person used less than 3/4 cup of paint.

11 Answer: 5

Explanation: Count the number of data points to the left of 1. Five 5 people used less than 1 cup of paint.

12 Answer: 30/360, 3/36, or 1/12

Explanation: A 30-degree angle turns through 30/360 of the circle.

13 Answer: C

Explanation: The angle is approximately half of right angle and turns through about 1/8 of the circle. The fraction 45/360 represents this value.

14 Answer: 45

Explanation: The first angle has a measure of 15 degrees. Aziz draws a total of 3 angles of that size. They have a combined measure of 15x3 or 45 degrees.

15 Answer: 720

Explanation: The wheel turns 360 degrees in 1 second. The wheel will turn 360x2 or 720 degrees in 2 seconds.

16 Answer:

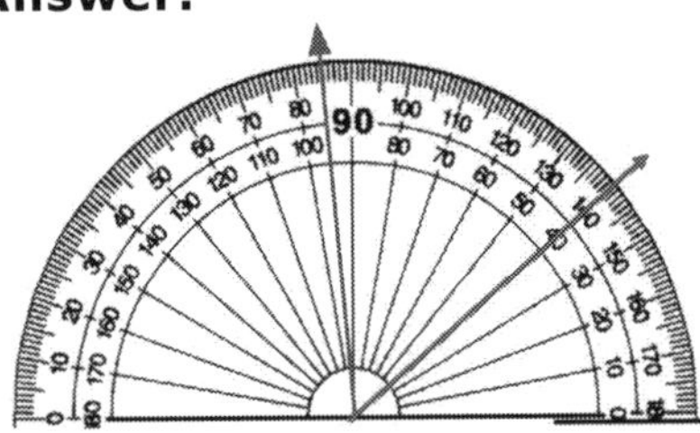

Explanation: The first ray is at the 40-degree mark (or 140-degree mark). Using the markings on the inside of the protractor, add 40 and 55 to determine the angle of the second ray, the 95-degree marking.
Or using the markings on the outside of the protractor, subtract 55 from 140 to determine the angle of the second ray, the 85-degree marking.

17 Answer: 126 – 26, 26 + 100, 154 – 54, 54 – 100

Explanation: Using the inside scale of the protractor, notice that the 2 rays are at the 26 and 126-degree markings. Using the outside scale of the protractor, notice that the 2 rays are at the 54 and 154-degree markings.

18 Answer: No

Explanation: The measure of Angle WXZ is 110 degrees. Write an equation: 110 = 42 + 4(6x3). Simplifying the right side gives 114. Thus, the measure is not 110 degrees.

19 Answer: 79

Explanation: calculate 15 + (7x2) + 100/2. Following correct order of operations, the expression changes to 15 + 14 + 50. Combine the terms with a result of 79.

20 Answer: 52

Explanation: The measure of Angle DEF is 120. The measures of the 2 smaller angles add to 68 (34 + 34 = 68). Subtract 68 from 120.

GEOMETRY
UNIT 1: LINES AND ANGLES

1 Answer: A

Explanation: A line goes without end in 2 directions.

2 Answer: C

Explanation: A line segment has 2 endpoints.

3 Answer: B

Explanation: A ray has 1 endpoint and goes without end in 1 direction.

4 Answer: A

Explanation: A line goes without end in 2 directions.

5 Answer: D

Explanation: A point in geometry is a location. It has no width, no length and no depth. A point is shown by a dot.

6 Answer: B

Explanation: An angle is formed by two intersecting rays that share a common endpoint where they meet.

7 Answer: C

Explanation: A line segment has 2 endpoints.

8 Answer: B

Explanation: A ray has 1 endpoint and goes without end in 1 direction.

9 Answer: D

Explanation: A line segment has 2 endpoints.

10 Answer: C

Explanation: Parallel lines stay the same distance apart. They will never intersect.

11 Answer: B

Explanation: Perpendicular lines are 2 lines which intersect at a right angle (a 90-degree angle).

12 Answer: B

Explanation: Parallel lines stay the same distance apart. They will never intersect.

13 Answer: A

Explanation: Parallel lines stay the same distance apart. They will never intersect.

14 Answer: B

Explanation: No angle has a measure of 90°. The triangle does not contain a right angle. The triangle is an obtuse triangle. One angle is greater than 90°.

15 Answer: A

Explanation: All three angles are less than 90°.

16 Answer: B

Explanation: All three angles are less than 90°. An obtuse triangle has one angle that is greater than 90°.

17 Answer: B

Explanation: The triangle is a right triangle, one angle is exactly 90°.

18 Answer: A

Explanation: The opposite sides will never cross either. So, these opposite sides are parallel.

19 Answer: B

Explanation: One pair of opposite sides will never cross. The shape has 1 pair of parallel sides.

20 Answer: B

Explanation: Perpendicular lines meet at right angles. They make square corners.

GEOMETRY
UNIT 2: TWO-DIMENSIONAL SHAPES

1 Answer: C

Explanation: The shape has four sides, so it is a quadrilateral. More specifically, the shape is a trapezoid because it has one pair of parallel sides..

2 Answer: D

Explanation: Four angles in the quadrilateral are marked as right angles.

3 Answer: C

Explanation: The shape has four sides of equal length, as shown by the markings, but the angles are not marked as right angles.

4 Answer: A

Explanation: The shape has four sides. It is a quadrilateral.

5 Answer: D

Explanation: A parallelogram has 2 pairs of opposite sides that have the same length. The markings show that the opposite sides have the same length.

6 Answer: C

Explanation: A rectangle is a quadrilateral with four right angles.

7 Answer: A

Explanation: The shape has four sides of equal length and four right angles, making it a square.

8 Answer: D

Explanation: The shape has four sides. It is a quadrilateral.

9 Answer: B

Explanation: A trapezoid is a quadrilateral with one pair of parallel sides.

10 Answer: A

Explanation: The triangle has 3 equal sides. It is an equilateral triangle.

11 Answer: C

Explanation: All sides are a different length and all angles are a different measure. This describes a scalene triangle.

12 Answer: D

Explanation: The triangle has 1 right angle.

13 Answer: A

Explanation: An isosceles triangle has 2 sides with the same length. It also has base angles with the same measure. The markings show that two sides have the same length.

14 Answer: B

Explanation: A trapezoid is a quadrilateral with one pair of parallel sides.

15 Answer: A

Explanation: A rectangle is a quadrilateral with four right angles.

16 Answer: B

Explanation: A rhombus is a quadrilateral whose sides are all the same length.

17 Answer: C

Explanation: Perpendicular line segments are line segments which intersect at a right angle.

18 Answer: A

Explanation: A quadrilateral is a polygon with 4 sides.

19 Answer: A

Explanation: A rectangle is a quadrilateral with four right angles.

20 Answer: A

Explanation: A quadrilateral is a polygon with 4 sides.

GEOMETRY
UNIT 3: LINE OF SYMMETRY

1 Answer: A

Explanation: A line of symmetry divides a shape into two regions that are mirror images of each other across the line.

2 Answer: B

Explanation: A line of symmetry divides a shape into two regions that are mirror images of each other across the line. The line does not divide the flower into 2 mirror images.

3 Answer: B

Explanation: A line of symmetry divides a shape into two regions that are mirror images of each other across the line. The figure does not have any lines of symmetry.

4 Answer: A

Explanation: A line of symmetry divides a shape into two regions that are mirror images of each other across the line.

5 Answer: A

Explanation: A line of symmetry divides a shape into two regions that are mirror images of each other across the line.

6 Answer: B

Explanation: A line of symmetry divides a shape into two regions that are mirror images of each other across the line. The line does not divide the flower into 2 two mirror images.

7 Answer: A

Explanation: A line of symmetry divides a shape into two regions that are mirror images of each other across the line.

8 Answer: B

Explanation: A line of symmetry divides a shape into two regions that are mirror images of each other across the line. The figure does not have any lines of symmetry.

9 Answer: B

Explanation: A line of symmetry divides a shape into two regions that are mirror images of each other across the line. The line does not divide the flower into 2 mirror images.

10 Answer: A

Explanation: A line of symmetry divides a shape into two regions that are mirror images of each other across the line.

11 Answer: B

Explanation: The capital letter "H" has a vertical line of symmetry and a horizontal line of symmetry.

12 Answer: A

Explanation: The letter "C" has a horizontal line of symmetry.

13 Answer: B

Explanation: A line of symmetry divides a shape into two regions that are mirror images of each other across the line. The line does not divide the flower into 2 equal parts.

14 Answer: A

Explanation: If a line were drawn from the center of the ladybug's head across center of it's back, it creates two mirror images.

15 Answer: B

Explanation: The image has 8 different lines of symmetry.

16 Answer: B

Explanation: The picture has no lines of symmetry.

17 Answer: B

Explanation: A vertical line of symmetry can be drawn downward at the center of the rainbow.

18 Answer:

Explanation: The figure has one vertical line of symmetry separating it into a part on the left that is a mirror image of the part on the right.

19 Answer:

Explanation: The figure has one vertical line of symmetry separating it into a part on the left that is a mirror image of the part on the right.

20 Answer:

Explanation: The figure has one vertical line of symmetry separating it into a part on the left that is a mirror image of the part on the right.

GEOMETRY - CHAPTER REVIEW

1 Answer: acute

Explanation: Angle, all 3 angles have measures that are less than 90 degrees.

2 Answer: right

Explanation: The triangle has one angle is exactly 90 degrees.

3 Answer: obtuse

Explanation: An obtuse angle has a measure that is more than 90 degrees, but less than 180 degrees.

4 Answer: acute

Explanation: An acute angle has a measure that is less than 90 degrees, but more than 0 degrees.

5 Answer: right.

Explanation: A right-angle has a measure of exactly 90°.

6 Answer: right

Explanation: A right-angle has a measure of exactly 90°.

7 Answer: straight

Explanation: A straight angle has a measure of exactly 180° and looks like a line.

8 Answer: B

Explanation: A rhombus is a parallelogram with all sides having the same length. The markings in this figure show that lines are parallel, when they are not parallel.

9 Answer: B

Explanation: None of the angles have a measure that is exactly 90-degrees.

10 Answer: A

Explanation: In an acute triangle, the measures of all three angles are less than 90°.

11 Answer: A

Explanation: A trapezoid is a quadrilateral with one pair of parallel sides. The markings show that the top side is parallel to the bottom side.

12 Answer: B

Explanation: The shape is a quadrilateral, but not a trapezoid since it does not have a pair of parallel sides.

13 Answer: B

Explanation: The shape is a quadrilateral, but not a trapezoid since it has no parallel sides.

14 Answer: B

Explanation: The shape is a quadrilateral, but might not be a trapezoid since it does not have markings showing one pair of parallel sides.

15 Answer: A

Explanation: The capital letter "E" has 1 horizontal line of symmetry.

16 Answer: B

Explanation: The letter "V" has 1 vertical line of symmetry.

17 Answer: B

Explanation: The shape has a vertical and a horizontal line of symmetry.

18 Answer: D

Explanation: The capital letter "R" has no lines of symmetry.

19 Answer: A

Explanation: The capital letter "Y" has 1 vertical line of symmetry.

20 Answer: False

Explanation: The image has 4 lines of symmetry.

GEOMETRY - EXTRA PRACTICE

1 Answer: 4

Explanation: The following pairs of rays are perpendicular: Ray BA and BD, Ray BD and BE, Ray BE and BG, Ray BG and BA.

2 Answer: A

Explanation: The shape has 6 angles in the interior. All of the interior angles have measures that are greater than 90 degrees and less than 180 degrees.

3 Answer: B

Explanation: When labeling an angle, the point or points used must show a distinct angle, with the vertex label in between the rays that form the angle.

4 Answer: 0

Explanation: There are no parallel lines in the shape. all sides will intersect if they are extended.

5 Answer: 13

Explanation: Each vertex of the polygon represents a corner on the polygon. The polygon has 13 corners.

6 Answer: rectangle, trapezoid, right triangle

Explanation: The two-dimensional shapes created by the lines in this picture include 2 triangles, 3 quadrilaterals (2 trapezoids and 1 rectangle).

7 Answer:

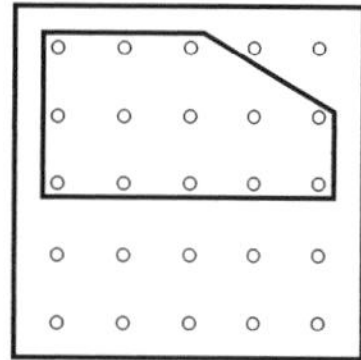

Explanation: Answers may vary. She can create a pentagon by moving the center any side of the square to create another corner corner so the square changes to a five-sided figure.

8 Answer: B

Explanation: The shapes in Group B all have 4 sides.

9 Answer: Yes

Explanation: A right triangle is defined as a triangle where at least one angle is a right angle. The remaining two angles have a sum of 90 degrees, which means they must both be acute angles.

10 Answer: Yes

Explanation: The triangle could be defined as isosceles and equilateral. It is isosceles because (at least) two sides have the same length. A more precise term would be equilateral, since the marks show that all sides have the same length.

11 Answer: C

Explanation: The letter "T" is an enclosed shape with 8 straight sidesshape (polygon). Octagons are polygons with 8 sides.

12 Answer:

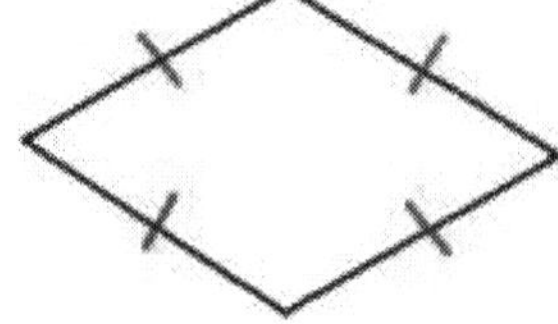

Explanation: The shape is a rhombus. A rhombus is a quadrilateral (specifically a parallelogram) with opposite equal acute angles, opposite equal obtuse angles, and four equal sides.

13 Answer: Yes

Explanation: The hexagon has two adjacent right interior angles, and six sides.

14 Answer: C

Explanation: The state of Utah is in the shape of a hexagon (has 6 sides). The states of Wyoming and Colorado border this state.

15 Answer:

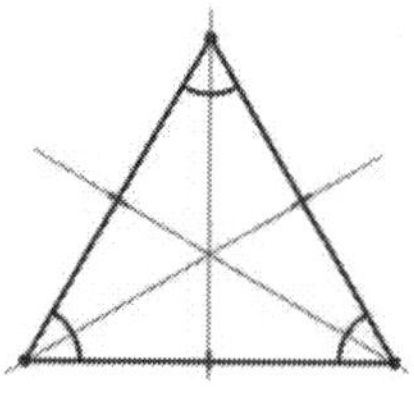

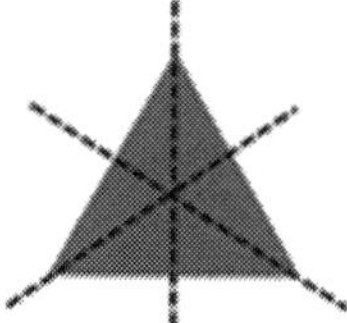

Explanation: A line of symmetry divides a figure into congruent parts that are mirror images of each other. An equilateral triangle has exactly 3 lines of symmetry.

16 Answer:

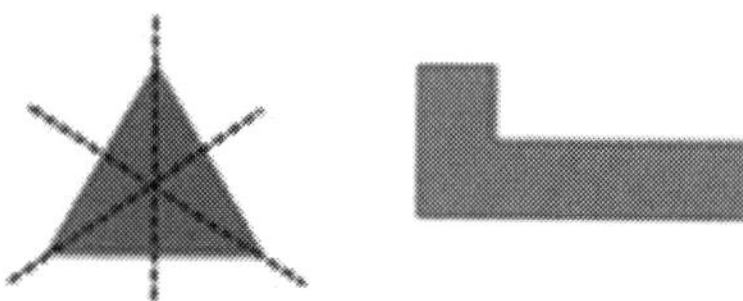

Explanation: The equilateral triangle has exactly 3 lines of symmetry. The hexagon (6-sided shape) is not a regular hexagon, and it has 0 lines of symmetry.

17 Answer: No

Explanation: While the leaf does appear to be similar on both sides of the line, the two halves of the leaf are not identical nor mirror images of each other.

18 Answer: D

Explanation: This model has 1 vertical and 1 horizontal line of symmetry.

19 Answer: No

Explanation: A non-regular quadrilateral is a good counter example. A quadrilateral is a shape with 4 sides. Not all quadrilaterals are symmetrical. Some trapezoids, for example, are not symmetrical.

20 Answer: circle

Explanation: A circle is an example of a two-dimensional shape that has an infinite number of lines of symmetry.

COMPREHENSIVE ASSESSMENTS
ASSESSMENT 1

1 Answer: No

Explanation: 5 times 6 means 5 groups of 6, not 5 plus 6.

2 Answer: Yes. There are 42 basketballs

Explanation: Multiplying 6 times 7 gives 42. There are 2 extra balls.

3 Answer: D

Explanation: Calculate $7 + 14 = 21$. Then, $40 - 21 = 19$.

4 Answer: $2 \times 4 = 8$ or $4 \times 2 = 8$

Explanation: Two times 4 means that you should multiply 2 and 4, which equals 8.

5 Answer: Agree

Explanation: All even numbers can be divided by two, so they are not prime, they all have 2 as a factor.

6 Answer: Three rows of 1 circle, 2 squares, 1 pentagon, 1 circle, 2 squares

Explanation: Answers must include a reasonable explanation such as, "Three horizontal rows of Three rows of 1 circle, 2 squares, 1 pentagon, 1 circle, 2 squares."

7 Answer: A

Explanation: The number 7 times 4 equals 28.

8 Answer: D

Explanation: Calculate the total number of points by multiplying 9×21, which is 189. Point V is slightly less than 200 on the number line.

9 Answer: 300,000

Explanation: Ten times 30,000 moves the decimal place in 30,000 one digit to the right from thirty-thousand to three-hundred thousand.

10 Answer: 8

Explanation: Calculate $11 \times 12 = 132$. Then, $132 - 124 = 8$.

11 Answer: B

Explanation: Partition the given number in the stem, 64, 072, into fourths using a variety of methods. For example, divide 64,072 by 4 which gives 16,018. Then, multiply the result by 3 to find the number of people who are older than 18 years old. This results in 48,054. Write the answer using words.

12 Answer: $8{,}000{,}000 + 400{,}000 + 70{,}000 + 6{,}000 + 5$

Explanation: Expanding a number requires rewriting the number into a sum of every digit shown as its place value.

13 Answer: C

Explanation: The value of the instruments (in order) are $5,000, $2,000, $2,000, and $4,000. Find the sum.

14 Answer: The number could be 8,914; 8,941; 9,481; 9,148; 9,184 or 9,418.

Explanation: Point A is at approximately 9,000. The given numbers must be arranged into a number that rounds to the nearest thousand as 9,000.

15 Answer: Yes.

Explanation: She correctly uses the "equal-addition" algorithm, which is adding tens to both the tens and ones before subtracting.

16 Answer: A

Explanation: A round trip is 1,118 miles each way, which is 2,236 miles. There are 4 round trips - multiply 2,236 by 4.

17 Answer: 2,789 x 4 = 11,156 miles

Explanation: The airplane makes a total of 4 trips over a 4-day period. Each trip is 2,789 miles.

18 Answer: C

Explanation: Determine the mph by dividing 441 by 7.

19 Answer: 1,704

Explanation: If the quotient is 213 and the divisor is 8, multiply 213x8.

20 Answer: C

Explanation: If 4 out of 24 socks are white and the rest are black, subtract 24 – 4 = 20 black socks. Thus, 20 pairs out of 24 pairs are black, which is equivalent to $\frac{10}{12}$ and $\frac{5}{6}$.

$\frac{20}{24} \div \frac{2}{2} = \frac{10}{12}$

$\frac{20}{24} \div \frac{4}{4} = \frac{5}{6}$

21 Answer: B

Explanation: The fraction $\frac{6}{8}$ is equivalent to $\frac{3}{4}$ or 0.75. The fraction $\frac{4}{5}$ is larger and is equivalent to 0.80. Thus, $\frac{4}{5}$ is the larger fraction.

22 Answer: $\frac{5}{8}$

Explanation: Subtract $\frac{2}{8}$ from $\frac{7}{8}$.

23 Answer: $\frac{6}{5}$ cups of flour; $\frac{2}{5}$ cups of sugar

Explanation: Doubling the recipe means multiplying the ingredients by 2. When multiplying fractions, multiply the numerator by 2, but do not change the denominator.

24 Answer: $1\frac{9}{10}$ or $1\frac{90}{100}$

Explanation: Add the two thicknesses: $\frac{40}{100} + 1\frac{9}{10} = 1\frac{9}{10}$ or $1\frac{90}{100}$.

25 Answer: Yes

Explanation: The difference between Point A and Point C is $\frac{6}{10}$ or $\frac{60}{100}$. Subtract $1\frac{9}{10} - 1\frac{9}{10} = \frac{6}{10}$ or $\frac{60}{100}$.

26 Answer: C

Explanation: Half of a dollar is 50 cents ($$0.50). Three tenths of a dollar is 30 cents ($0.30).

27 Answer: B

Explanation: Four-hundredths of a dollar is 4 cents ($0.04), and eight tenths of a dollar is 80 cents ($0.80).

28 Answer: B

Explanation: Iris has $0.28, and Joaquin has $0.22. The value $0.28 is greater than $0.22.

29 Answer: C

Explanation: When comparing decimals, compare digits beginning on the left side. The decimal 0.3 has one less tenth than 0.46.

30 Answer: B

Explanation: Intersecting lines meet at one point. The given lines do not intersect.

31 Answer: B

Explanation: A quadrilateral is a four-sided shape. A trapezoid is a quadrilateral with one pair of parallel sides.

32 Answer: B

Explanation: The soccer ball appears to have a pentagonal pattern and has more 5 lines of symmetry.

33 Answer: 20 mm

Explanation: 1 centimeter = 10 millimeters

34 Answer: 1,467 grams

Explanation: 1 kilogram = 1,000 grams

35 Answer: B

Explanation: One tablespoon is equivalent to 3 teaspoons.

36 Answer: C

Explanation: One cup is equivalent to 16 tablespoons, so $\frac{1}{4}$ cub is equivalent to 4 tablespoons.

37 Answer: A

Explanation: Add all the side lengths.

38 Answer: A

Explanation: Multiply the side length by itself: $10 \times 10 = 100$

39 Answer: 5

Explanation: Count the number of data points for $\frac{1}{2}$ inch and for 1 inch and subtract.

40 Answer: B

Explanation: The angle turns entirely around a circle. It has a measure of 360 degrees.

41 Answer:

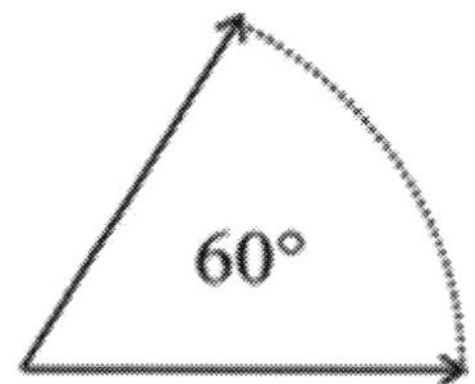

Explanation: Robin turns the plant 10x6 degrees, or 60 degrees.

42 Answer: D

Explanation: The measure of Angle ABE is 170 degrees. Angle ABC is equivalent to 170/2, or 85 degrees.

43 Answer:

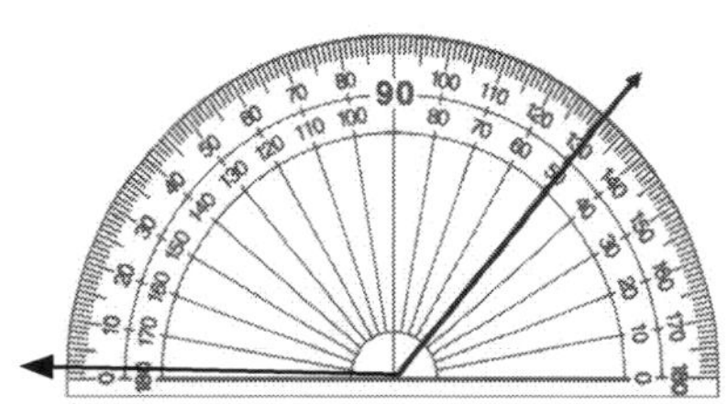

Explanation: Using the inside scale – the first ray is at 50 degrees. Add 50 + 128 to determine the second ray should cross through the 178-degree marking.
Using the outer scale – the first ray is at 130 degrees. Subtract 128 from 130 to determine the second ray should cross through the 2-degree marking.

44 Answer: 69 degrees

Explanation: Adjacent angles share a vertex and a side, but no interior points. Add the measures of adjacent angles to find the total measure: $43 + 26 = 69$.

45 Answer: 124 degrees

Explanation: The measure of the angle can be determined using this expression: $31 + (31 \times 3)$.

COMPREHENSIVE ASSESSMENTS ASSESSMENT 2

1 Answer: 28 x 3 = p
p = 84

Explanation: The equation shows 28 groups of 3, which finds how many pencils there are in all. There are 84 pencils in all.

2 Answer: 25

Explanation: Multiply 3 times 25 equals 75 or divide 75 by 3 equals 25.

3 Answer: 3

Explanation: There are 48 pencils, $\frac{48}{9} = 5$ with a remainder of 3. there are 3 pencils left over.

4 Answer: 7 x 6 = 42

Explanation: Answers must include a reasonable explanation such as, "This story can be written as 7 x 6 = x because 7 groups of 6 is the same as 7 times 6.

5 Answer: No

Explanation: The number 3 is not a factor of 76. Dividing 76 by 3 leaves a remainder. Thus, 3 cannot be multiplied by another whole number to produce 76.

6 Answer: 35

Explanation: If the rule is "add 6", then subtract 6 to find numbers in the front of the sequence. Thus, 53 minus 6 is 47, 47 minus 6 is 41, and 41 minus 6 is 35. The sequence is 35, 41, 47, 53... .

7 Answer: D

Explanation: To find the maximum, multiply,5 times 29 which equals 145.

8 Answer: Eric: 17,000
Justin: 1,700
Veronica: 17
Brandy: 2,017

Explanation: Eric's number is given. To find Justin's number divide 17,000 by 10. To find Veronica's number, divide Justin's number by 100. To find Brandy's number, add 2,000 to the Veronica's number.

9 Answer: Tara

Explanation: Tara spends 10 times as much as Sarah, and 100 times as much as Monique.

10 Answer: 25

Explanation: If half of the fish are blue, divide 80 by 2, giving 40 blue fish. Then, subtract 80 – 40 – 15 = 25. Of the fish in the tank, 25 are yellow.

11 Answer: D

Explanation: Subtracting 300,019 and 64, 105 will result in 235,914. Expand this number.

12 Answer: Nine million one hundred forty-seven thousand one hundred twenty

Explanation: Writing a number in written form is an expression of the number following base-ten numeral conventions.

13 Answer: 2,537

Explanation: The number rounds to 2,500. The only combination of the given numbers that rounds correctly is 2,537.

14 Answer: 13,810
14,495

Explanation: The number 13,810 rounds to 13,800, and the number 14,495 rounds to 14,500.

15 Answer: Explain the regrouping necessary for the ones, tens, hundreds and thousands place.

Explanation: Sample Response: Regroup 1 ten as 10 ones, then subtract 9 ones from the 10 ones, 10 – 9 = 1.
Regroup 1 ten thousand as 10 thousands, then regroup 1 thousand as 10 hundreds, and 1 hundred as 10 tens.
Subtract the tens, 120 – 50 = 70
Subtract the hundreds. 900 – 100 = 800
Subtract the thousands. 9,000 – 5,000 = 4,000
Add the partial sums.
4,000 + 800 + 70 + 1 = 4,871

16 Answer: Both strategies use decomposition of numbers.

Explanation: Strategy A decomposes 11 into 10 and 1 before multiplying. Strategy B recognizes that multiplying 35 by 11 is the same as multiplying 35 by 10 and then adding an additional 35 to that product.

17 Answer: 640

Explanation: Find the total number of quarters by multiplying 16 x 40.

18 Answer: No

Explanation: Lyndon used the partial quotient strategy to divide multi-digit numbers, but the place value relationships must be preserved. In rounding 450 to 480, a number which is compatible with 8, he should recognize the quotient of 450 divided by 8 is less than 60. In rounding 450 to 400, a different compatible number, the quotient is 50. This means 450 divided by 8 is between 50 and 60.

19 Answer: C

Explanation: The product (or answer to the multiplication problem) can be found by adding all the products inside of the model. The products are 20x10 =200, 20x7 = 140, 5x10 = 50, and 5x7 = 35.

20 Answer: A

Explanation: The fraction of cards that are soccer cards is $\frac{15}{20}$ because 5 are baseball cards and the rest are soccer cards, $20-5=15$. Then, 15 out of 20 cards are soccer cards. The fraction $\frac{15}{20}=\frac{3}{4}$ because $\frac{15}{20}\div\frac{5}{5}=\frac{3}{4}$.

21 Answer: B

Explanation: The fraction $\frac{6}{7}$ is closest to 1. The fraction $\frac{3}{8}$ is close to $\frac{1}{2}$. The fraction $\frac{1}{4}$ is closest to 0.

22 Answer: $\frac{1}{6}+\frac{1}{6}+\frac{1}{6}+\frac{1}{6}$

Explanation: Altogether, $\frac{4}{6}$ of the pie was eaten. This can represented by adding $\frac{1}{6}$ to itself 4 times.

23 Answer: Yes

Explanation: Three-fourths is a multiple of one-fourth. Then, 8 groups of $\frac{3}{4}$ is equivalent to 24 groups of $\frac{1}{4}$.

24 Answer: 10

Explanation: She adds the fraction, $\frac{3}{10}$, 10 times.

25 Answer: No

Explanation: The number of blue and green marbles in the bag is $\frac{40}{100}$ or $\frac{4}{10}$, which is less than half of the bag.

26 Answer: C

Explanation: Itzel has 34 cents, and Henrietta has 40 cents. The difference is 6 cents, which is expressed as $0.06 in decimal form.

27 Answer: B

Explanation: The mixed number , read as "Five and seven-tenths", is expressed as a whole number plus a decimal with one digit in the tenths place. This decimal is 5.7.

28 Answer: 0.27

Explanation: There are 23 squares shaded, which represents 0.23. Half of the figure is represented as 0.50. The difference between 0.50 and 0.23 is 0.27.

29 Answer: No

Explanation: Although 0.60 is greater than 0.50, the size of the first pie is much smaller than the size of the second pie.

30 Answer: A

Explanation: Intersecting lines meet at a point.

31 Answer: right

Explanation: In a right triangle, one angle is exactly 90°.

32 Answer: 2

Explanation: The figure has a horizontal and vertical line of symmetry.

33 Answer: B

Explanation: 1 liter = 1000 milliliters

34 Answer: B

Explanation: 1 meter = 1000 millimeters

35 Answer: 11,300

Explanation: There are 1,000 meters in 1 kilometer.

36 Answer: 9,150

Explanation: There are 1,000 meters in 1 kilometer.

37 Answer: 7 x 4 or 4 x 7

Explanation: The problem says that the blue rectangle has an area of 28, so list dimensions of a rectangle that give an area of 28 and a perimeter of 22. The whole numbers that divides evenly into 28 are 4 and 7. The perimeter is $2(4) + 2(7) = 22$.

38 Answer: 180

Explanation: The perimeter is 36 meters. Multiply the price per meter by the perimeter.

39 Answer: 1

Explanation: Count the number of data points to the right of $\frac{1}{2}$.

40 Answer: A

Explanation: The size of the angle is less than a right angle, but greater than one-half of a right angle, which is $\frac{70}{360}$ of the circle.

41 Answer: 35

Explanation: An angle that turns through 35 1-degree angles has a measure of 35 degrees.

42 Answer: 105

Explanation: The measure of angle ABF is 35 degrees. The measure of angle ABC is 35×3 or 105 degrees.

43 Answer: B

Explanation: Find the measure of the angle by subtracting 22 from 131.

44 Answer: B

Explanation: Adjacent angles share a vertex and a side. The measures of the two angle add up to 69, so $69 - 42 = 27$.

45 Answer: 110

Explanation: The straight angle is 180 degrees. The two unmarked angles have a measure of $180 - (30 + 40) = 110$.

72888666R00137

Made in the
USA
Middletown, DE